Synthetic & Renewable Fuels 2021

Topical Conference at the 2021 AIChE Annual Meeting

Boston, Massachusetts, USA and Online
7 - 11 November and 15 - 19 November 2021

ISBN: 978-1-7138-5695-5

Printed from e-media with permission by:

Curran Associates, Inc.
57 Morehouse Lane
Red Hook, NY 12571

AIChE
120 Wall Street, FL 23
New York, NY 10005-4020

Phone: (800) 242-4363
Fax: (203) 775-5177

www.aiche.org

Additional copies of this publication are available from:

Curran Associates, Inc.
57 Morehouse Lane
Red Hook, NY 12571 USA
Phone: 845-758-0400
Fax: 845-758-2633
Email: curran@proceedings.com
Web: www.proceedings.com

TABLE OF CONTENTS

Chemical upcycling of waste polyolefinic plastics to low-carbon synthetic naphtha for closing the plastic use loop

Leilei Dai*, Nan Zhou, Paul Chen, Kirk Cobb, Roger Ruan*

Center for Biorefining and Department of Bioproducts and Biosystems Engineering
University of Minnesota, 1390 Eckles Avenue, St. Paul, MN 55108, USA

dai00087@umn.edu (L. Dai); ruanx001@umn.edu (R. Ruan)

ABSTRACT

Plastic materials are extremely popular all over the world and used in the different walks of life due to the inherent properties of being strong, lightweight, and easily shaped. However, the vast majority of waste plastics ever produced enters into landfills or our ecosystems, creating a plastic waste crisis. Polyolefins, such as high density polyethylene (HDPE), low density polyethylene (LDPE), and propylene (PP), represents the largest amount of discarded waste plastics, accounting for over half of the waste plastic stream.

Currently, the most common solution to recycle the plastic waste is mechanical method that is limited by the corresponding challenges. The recycled plastics after melting and remolding generally have poorer properties than those of the virgin plastics, enabling the recycled plastics to be only applicable to lower quality products. This downcycling process makes the waste plastic recovery not attractive in the industry. Developing an effective pathway to remove waste plastics from landfill and incineration plants and create a circular economy requires a more appropriate technology beyond the conventional mechanical recycling by melting and re-molding. Pyrolysis has shown the great potential of plastic recycling for energy, fuels, chemicals, and materials production, enabling the plastic wastes to stay in the economy and out of the environment. Recently, we have screened different catalysts to improve the yield and quality of the liquid hydrocarbons from plastic waste pyrolysis, with the aim of maximizing naphtha fractions for new plastic production. Notably, relay catalysis of Al_2O_3 followed by ZSM-5 achieved up to 100% conversion into monoaromatics and C5-C12 alkanes/olefins at a catalyst to plastic ratio of 4:1. In order to further mitigate the aromatic formation, Al_2O_3 pillared clay catalyst was developed and tested, resulting over 50% C5-C12 alkanes and less than 10% monoaromatics. The liquid yield can reach over 70 wt.% with 52% of C5-C12 alkanes, 17% of C5-C12 olefins, 14% of C5-C12 aromatics, and 17% of C13+ hydrocarbons from catalytic pyrolysis of polyethylene. More importantly, this process is also suitable for waste plastic mixtures, e.g., LDPE 41%, HDPE 24%, and PP 35%, which represents one of the typical plastic waste streams. Furthermore, we proposed a novel approach (**Figure 1**) of combining two catalytic reforming zones, where the first catalytic reforming zone was intended to improve the cracking of polyolefins into short chain olefins, and the second, lower-temperature catalytic reforming zone was intended for the hydrogenation process to convert these olefins to C5-C12 alkenes. The tests were very successful and very promising results (60-75% C5-C12 alkanes, 3-5% C5-C12 olefins, 5-15% mono-aromatics) were obtained as hypothesized.

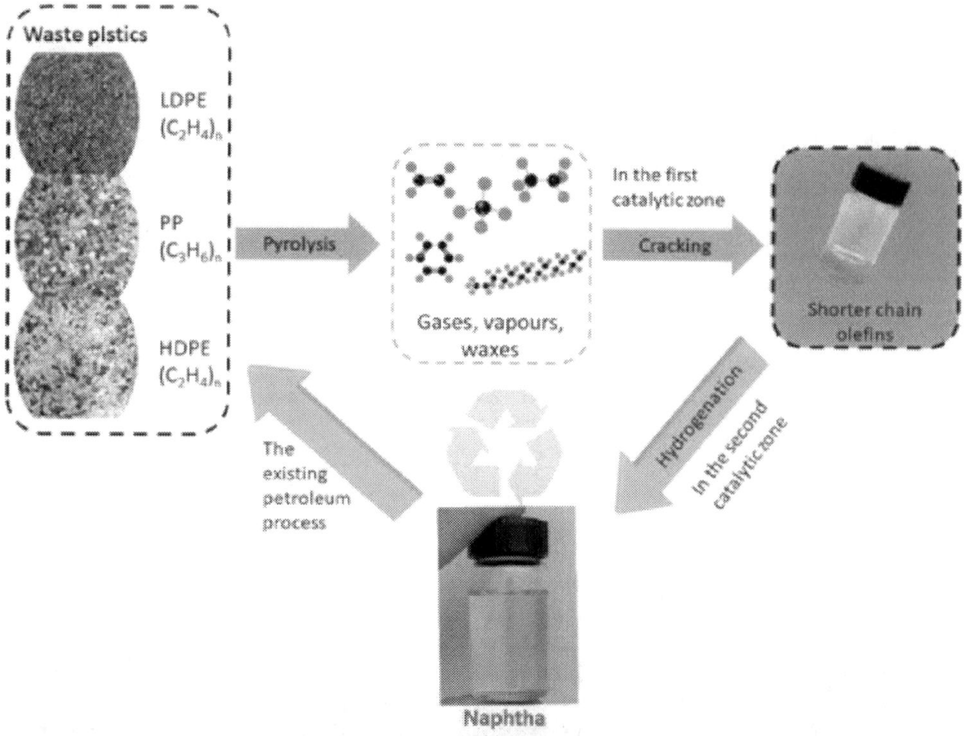

Figure 1. Converting polyolefinic plastics to high quality naphtha via pyrolysis coupled with a tandem catalysis process.

Table 1 shows the mass balance of the plastic-to-naphtha technology. When just using one catalytic zone with Zn/SBA-15 as a catalyst, 51.67 wt.% of liquid oil can be obtained, which is higher than that of the two catalytic zone (36.65 wt.%). It may be because the second catalytic zone was operated at lower temperature and some unconverted long chain hydrocarbons with high boiling temperature were trapped in this zone. This also highlights the importance of wax cracking in the first catalytic zone. It can be noted that after adding the second catalytic zone, hydrogen content in the gas product experiences a decline. That supports our hypothesis that the olefins produced in the first catalytic zone will be converted into paraffins by consuming the spontaneous hydrogen or hydrogen transfer reaction. The liquid oil composition verifies that the tests were very successful and very promising results were obtained as hypothesized.

Table 1. Mass balance for the catalytic reforming of HDPE pyrolysis volatiles.

	One catalytic zone	Two catalytic zone
Product yields (wt.%)		
Gas yield	34.99	30.60
Liquid oil yield	51.67	36.65
Coke on the first catalytic zone	13.34	15.05
Coke on the second catalytic zone		17.70
Chemical composition of products		
Dry gas (vol.%)		
Hydrogen	25.97	17.62
Methane	11.72	21.94
Ethylene	9.72	23.44
Ethane	15.04	12.27
Liquefied petroleum gas (vol.%)		
Propylene	9.07	13.43

Propane	9.36	6.80
Butylene	4.61	12.21
Butane	1.68	5.12
Liquid oil (%)		
C5-C12 n-paraffins	5.42	17.61
C5-C12 iso-paraffins	2.41	41.89
C5-C12 olefins	22.49	4.78
C5-C12 naphthene		7.83
C5-C12 aromatics	4.66	10.56
C13-C22 hydrocarbons	63.29	17.33
C23-C30 hydrocarbons	1.73	

Reaction conditions: Pyrolysis temperature = 500 °C, weight of HDPE, Zn/SBA-15, and Al2O3 pillared Montmorillonite clay = 6, 3, and 15 g, first catalytic zone temperature = 500 °C, second catalytic zone temperature = 250 °C, flow rate of nitrogen = 50 ml/min.

The promising results of the two catalytic zone is mainly attributed to the cracking and hydrogenation in the first and second catalytic zone, respectively (**Figure 2**). The catalytic activity of Al_2O_3 pillared M-clay may be due to the special features of this material, including the large pore size and Al_2O_3-induced active sites (more Brønsted acid sites). On one hand, the large pore size alleviates the steric and diffusional hindrances for entering the catalyst channels because the bulky nature of the polyolefins or long chain hydrocarbon intermediates in the pyrolysis reaction; on the other hand, Brønsted acid sites can be responsible for carbonium ion formation which is cracked into alkanes further. Based on the results above, it is supposed that this catalyst increases the share of C5-C12 alkanes by primarily facilitating the following reactions.

$$\underset{\text{Olefins}}{3C_nH_{2n}} + \underset{\text{Cyclo-alkanes (Olefins)}}{C_mH_{2m}} \longrightarrow \underset{\text{Alkanes}}{3C_nH_{2n+2}} + \underset{\text{Aromatics}}{C_mH_{2m-6}} \qquad \text{Hydrogen transfer reaction}$$

$$\underset{\text{Olefins}}{C_nH_{2n}} \xrightarrow{\text{H addition}} \underset{\text{Alkanes}}{C_nH_{2n+2}} \qquad \text{Hydrogenation}$$

In the HDPE degradation, the long polymer chain is initially cracked into shorter chain olefins, which would then be converted to alkanes and aromatics via the hydrogen transfer reaction over Al2O3 pillared M-clay catalyst.

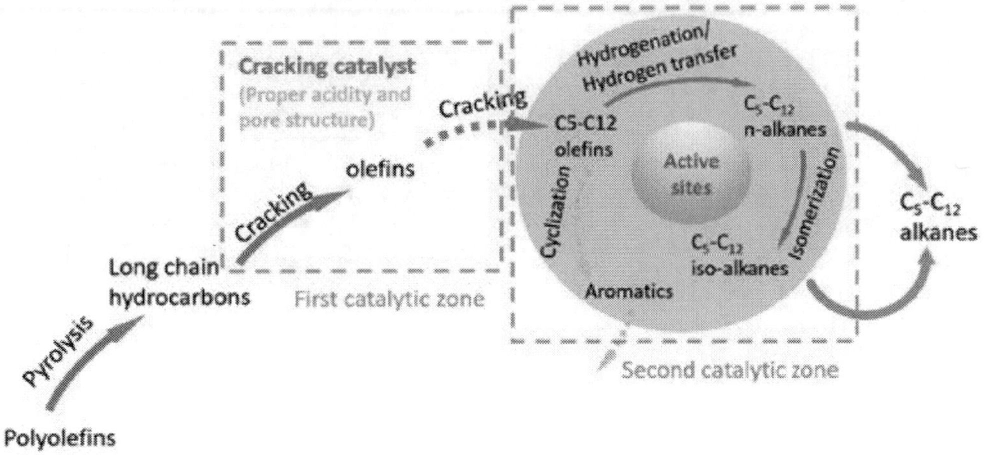

Figure 2. Reaction scheme for the two catalytic zones.

To advance this technology towards industrial applications, a continuous catalytic microwave-assisted pyrolysis (cMAP) system with a processing capacity of 1 ton/day was developed. The system features a mixing bed of silicon carbide balls enabling fast, uniform and energy-efficient heating for the process. Through this system, 57 wt.% C5-C22 liquid hydrocarbons was obtained from polyethylene, allowing over 32% energy saving compared to production of similar products from virgin materials.

KEYWORDS: Catalysis; Environmental sustainability; Chemical Reaction Engineering

631656 Deconstruction of Waste Polyethylene into Liquid Hydrocarbon Fuels and Lubricants over Ru Catalyst

Abstract:

Polyethylene (PE) is the most popular plastic globally, and the widespread use of plastics has created severe environmental issues. High energy consumption in the current process makes its recycling a challenging problem. In our report, the depolymerization of high-density polyethylene (HDPE) was conducted in various liquid-phase solvents with the Ru/C catalyst under relatively mild conditions. The maximum yields of the jet-fuel-range and lubricant hydrocarbons were 60.8 wt% and 31.6 wt%, respectively. After optimizing the reaction conditions (220 °C and 60 bar of H_2), the total yield of liquid hydrocarbon products reached approximately 90 wt% within only one hour. The product distribution could be tuned by the H_2 partial pressure, the active metal particle size, and the solvents. The solvation of PE in the different solvents determined the depolymerization reaction kinetics, which was confirmed by the molecular dynamics (MD) simulation results.

622570 Optimal Design of Controlled Environment Agricultural Systems Under Market Uncertainty

Abstract:

Global population growth and increased climate variability have posed major challenges to traditional agricultural production [2,6]. As these trends persist, meeting increased consumer demand under heightened uncertainty will further stress the capabilities of our existing food production infrastructure [4,5]. For this reason, alternative methods are being sought to augment conventional food production. Controlled environment agriculture (CEA) is an alternative strategy that decouples crop production from environmental conditions through indoor cultivation. CEA offers a more sustainable alternative to conventional agriculture and presents opportunities for improved food security by increasing public access to nutritious fresh produce. Startup costs for CEA systems, however, are significantly higher than those of traditional growing operations, and therefore present a larger barrier to adoption. To accelerate the adoption of this potentially disruptive technology for next generation food production, a better understanding of the long-term economic viability of CEA technology is desired. To provide this valuable insight, we have developed the first economics-based optimal design and planning model for CEA systems.

To date, the majority of agricultural planning models have focused on traditional outdoor food production. Many of the existing models have been tailored to highly specific operations, few have employed mathematically rigorous optimization techniques, and only a handful have accounted for uncertainty, which is desired to realistically assess the long-term economic viability of the system [1,3,7]. Since existing models are not applicable to CEA systems, we have developed an optimal design and planning model that can be used to assess the long-term economic viability of a broad range of CEA systems. Our proposed model employs robust optimization to account for uncertainty in the design and planning stages of CEA production, which enables risk-averse investment decision making for long-term economic viability.

In this work, we present a novel methodology for the simultaneous robust design and planning of CEA systems under multi-period market price uncertainty/risk. This problem is cast as a bilevel program with a nonconvex outer program pertaining to the design objective and multiple semidefinite inner programs pertaining to mean-variance risk exposure over each period considered in the planning horizon. The bilevel program is reformulated as a nonconvex semi-infinite program with several semi-infinite constraints pertaining to the multi-period risk exposure over the horizon. The problem is solved to global optimality using a custom branch-and-bound strategy constructed with EAGO.jl [10] that exploits a combination of McCormick-based envelopes and Gurobi's nonconvex solver [8], to solve the relaxed (upper-bounding in the maximization sense) subproblems, and IPOPT [9] for local solution of the lower-bounding (in the maximization sense) subproblems. A solution to the model represents a design and operating schedule that is robust to worst-case market uncertainty and therefore provides a conservative basis for CEA engineering and investment decision-making.

Our proposed model has been formulated generally and can be used to optimize large-scale CEA systems for the cultivation of any crop portfolio via single or multi-mode cultivation. The optimization objective employed is to maximize the net present value of the CEA system over its expected lifespan. Constraints limiting risk exposure and production capacity are also included to ensure that the optimal design is both robust and practical. Capital and operating expenses, including those associated with system construction, cultivation equipment, resources, labor, and distribution, are accounted for in the economic objective through sub-models that capture the location, capacity, and crop allocation cost dependencies of the specific CEA system being modeled. These sub-models can be easily adjusted for different CEA configurations without structural changes to the main optimization algorithm. This not only enables efficient determination of the optimal design for a given CEA configuration, but also enables efficient comparison between multiple candidate systems for strategic adoption of CEA technology.

To evaluate the efficacy of our approach, the economic performance of the robust design was compared against that of naïve design and operating strategies through two case studies. In each of the presented case studies, an economically feasible design was identified, even when the naïve design was economically infeasible, and the improvement in system performance attributed to the implementation of a robust design was quantified. This approach represents, to our knowledge, the first robust optimization approach to CEA systems and is demonstrated to effectively increase the robustness of CEA systems to market uncertainty, improve the long-term economics of CEA systems over naïve operating strategies, and validate the economic viability of single and multi-mode CEA production of distinct crop portfolios. This work establishes a means of evaluating the economic viability of a wide array of CEA systems and serves as a baseline for future development of CEA design and planning models to accelerate the adoption of CEA technology.

References:

[1] Ahumada, O., Villalobos, J.R., 2009. Application of planning models in the agri-food supply chain: a review. Eur. J. Oper. Res. 196 (1), 1–20. doi: 10.1016/j.ejor.2008.02.014 .

[2] Deschênes, O., Greenstone, M., 2007. The economic impacts of climate change: evidence from agricultural output and random fluctuations in weather. Am. Econ. Rev. 97 (1), 354–385. doi: 10.1257/aer.97.1.354 .

[3] Dury, J., Schaller, N., Garcia, F., Reynaud, A., Bergez, J.E., 2012. Models to support cropping plan and crop rotation decisions. A review. Agron. Sustain. Dev. 32 (2), 567–580. doi: 10.1007/s13593- 011- 0037- x .

[4] Fedoroff, N.V., 2015. Food in a future of 10 billion. Agric. Food Secur. 4 (1), 11. doi: 10.1186/s40066- 015- 0031- 7 .

[5] Field, C.B., Barros, V., Stocker, T.F., Dahe, Q., 2012. Managing the Risks of Extreme 1083 Events and Disasters to Advance Climate Change adaptation: Special Report of the Intergovernmental Panel on Climate Change (IPCC). Cambridge University Press.

[6] Food and Agriculture Organization of the United Nations (FAO), 2018. The state of agricultural commodity markets 2018: Agricultural Trade, Climate Change and Food Security. Technical Report. FAO. URL http://www.fao.org/3/I9542EN/i9542en.pdf .

[7] Glen, J.J., 1987. Mathematical models in farm planning: a survey. Oper. Res. 35 (5), 641–666. doi: 10.1287/opre.35.5.641 .

[8] Gurobi Optimization, LLC, 2020. Gurobi optimizer reference manual. URL http://www.gurobi.com .

[9] Wächter, A., Biegler, L.T., 2006. On the implementation of an interior-point filter line-search algorithm for large-scale nonlinear programming. Math. Program. 106 (1), 25–57. doi:

10.1007/s10107- 004- 0559- y.

[10] Wilhelm, M.E., Stuber, M.D., 2020. EAGO.jl: Easy advanced global optimization in Julia. Optim. Methods Softw. doi: 10.1080/10556788.2020.1786566 .

632232 Reproducible and Fast Mapping of Material Flows in Industrial Networks Using Piot Hub: A Novel Cloud Based Computational Tool

Abstract:

An important goal of industrial ecology is to ensure a balance is maintained between economic and environmental sustainability. Facilitation of this balance requires an understanding of the structure of the material economy and the material flows/waste generation in industrial networks in order to minimize the environmental consequences of these industrial systems. Physical Input-Output Tables (PIOTs) are uniquely suited for this purpose by mapping physical flows between industrial sectors. In previous work, PIOT-Hub was developed to automate the generation of both Physical Supply Tables (PSTs) and Physical Use Tables (PUTs) as well as their conversion into PIOTs by integrating the strength of mechanistic models and macroeconomic Input-Output framework [1]. Hence, PIOT-Hub, a cloud-based tool can be used to generate the PST and PUT for a region using a bottom-up approach where relevant industrial processes are simulated via mechanistic models to determine steady-state material input and output requirements. Thus, the tool establishes a connection between engineering and economic model in such a way that it is insightful for transitioning to a low carbon and zero waste circular economy. In past work, PIOT-Hub was applied to 11 agro-based economic sectors in Illinois, thus capturing elemental flows across the economy, demonstrating the automation aspect. Here we evaluate the reproducibility and time aspect of PIOT-Hub to map the material flows of equivalent economic sectors in Indiana. We develop simulation models for industries in Indiana and utilize PIOT-Hub for mapping the industrial network, thus evaluating the functionality of this novel computational platform. This work not only demonstrates the reproducibility of the tool's utilization to other geographic regions, but also serves as a case study of how the fast generation of PSTs, PUTs, and PIOTs by PIOT-Hub can be used to compare the material economies of multiple regions. We perform a comparative analysis of material flows in Illinois and Indiana using our work and the PIOT generated for Illinois in previous work. We will also demonstrate the collaborative aspect of this computational tool by live demo of the tool during this talk, thus highlighting the scaling impact of collaboration through cloud-based platform in making sustainable manufacturing decisions.

References

[1] Shweta Singh; Lan Zhao; Jaewoo Shin; Venkata Sai Gargeya Vunnava (2021), "PIOT hub for Automated Generation of Regional PIOTs," *https://mygeohub.org/resources/piotpy.*

625295 Solid Oxide Fuel Cell – Thermal Energy Storage (SOFC-TES) Integrated System Modeling and Energy Efficiency Optimization.

Abstract:

Concerns over greenhouse gas emissions and their associated environmental challenges continue to increase. As a result, cleaner energy generation technologies, such as fuel cells, are receiving attention as the next generation's energy sources. One of these attractive energy technologies is the solid oxide fuel cells (SOFCs) due to its several advantages such as 1) reduced greenhouse gas emissions, 2) higher efficiency, and 3) flexible fuel selection compared to other types of fuel cells. However, the temperature of flue gas from SOFCs is significantly high due to its operating temperature between 600°C and 1000°C. Therefore, it is necessary to recover this waste heat so that we can maximize the process energy efficiency.

Many studies have been conducted to utilize the high-temperature exhaust gases from SOFCs so far. However, they only focused on the direct utilization of the heat through the electricity generation, using processes such as gas turbine[1], Rankine cycle[2], and combined heat and power (CHP)[3] in order to maximize the process energy efficiency. However, the aforementioned utilization methods result in losses where the generated electricity is further converted to other forms of energy such as thermal energy for space heating. The energy losses during this process are often significant.

Motivated by this, we propose the SOFC-integrated thermal energy storage (TES) method as a more efficient utilization technique for recovering waste heat from SOFCs. We employed a 4 kW-SOFC and combined it with the thermal energy storage system for storing the excess energy from the high-temperature exhaust gas. In order to integrate both processes, SOFC and TES mathematical models were developed using MATLAB. Next, the developed models are integrated, simulated, and the overall efficiency of the SOFC-TES model is compared to the general utilization methods. To further improve the proposed model's thermal efficiency, we developed a heat exchanger network using Aspen Energy Analyzer. Through the developed heat exchanger network, the integrated model was optimized by pinch analysis.

The optimization results revealed that the process efficiency of the integrated system, that is, SOFC-TES, showed superiority compared with the traditional methods of utilizing the waste heat from SOFCs. By this, excess heat from the exhaust gases can be reliably stored during low-demand seasons and reutilized during high-demand seasons to supplement residential and industrial energy demand or deficits. We concluded that the SOFC-TES could be widely used for various purposes due to its inherent characteristics, i.e., storing heat, such as a greenhouse in the smart-farm and human life, compared with the traditional method of waste heat utilization from the SOFCs.

References

[1] S.H. Chan, H.K. Ho, Y. Tian, **Modeling of simple hybrid solid oxide fuel cell and gas turbine power plant**, Journal of Power Sources, 109 (1) (2002), pp. 111-120

[2] Masoud Rokni, **Thermodynamic analysis of an integrated solid oxide fuel cell cycle with a Rankine cycle**, Energy Conversion and Management, 51 (2010), pp. 2724-2732

[3] S.H. Chan, C.F. Low, O.L. Ding, **Energy and exergy analysis of simple solid-oxide fuel-cell power systems**, Journal of Power Sources, 103 (2002), pp. 188-200

630767 Upcycling Food Waste for Hydrogen Storage: Technoeconomic Assessment

Abstract:

While annually 1.6 billion tons of food waste is estimated to be discarded globally by 2030, according to U.S. Environmental Protection Agency (EPA) the corresponding figure is projected forecast to reach approximately 66 million tons for U.S. With the monumental challenge of food waste management, researchers are seeking avenues of upcycling FW into higher value materials such as hydrogen storage medium. Therefore, the objective of this research was to evaluate the technoeconomic prospect of activated carbon production from food waste for the ultimate purpose of hydrogen storage through the application of hydrothermal carbonization of food waste followed by chemical activation of the hydrochar using KOH. The process was simulated using Aspen Plus and the process flow diagram (PFD) was constructed such that it mainly constituted of 2 rotary kilns and filter press, 1 rotary cooler, dryer, homogenizer, as well as other necessary equipment - pumps, heat exchangers, storage tanks. The facility processes 96 ton of food waste per day to be converted to activated carbon powder with a carbon yield of 20.4%. Also, it was assumed that the activated carbon manufacturing facility obtains food waste of Brevard County, Florida and the facility operates continuously 330 days a year. Several parameters were incorporated in the study including equipment sizing, capital costs and operating costs, such as labor, utilities, maintenance and equipment depreciation. Break-even cost from the economic analysis was evaluated and sensitivity analysis was carried out to reveal the effect of 7 different process variables on the break-even cost where raw materials' (KOH and food waste) and wastewater treatment cost significantly affected it. The unprecedented analysis shed light on the economic feasibility of producing activated carbon from food waste as a cost-effective route of upcycling it.

625694 Economic and Environmental Impact of Biochar Addition on Anaerobic Digestion Process

Abstract:

Anaerobic digestion (AD) technology is becoming a promising pathway for producing clean and renewable natural gas from waste resources. Despite growing industrial adoption, AD technologies still suffer from inefficiency due to process instability and microbial inhibitions. Scientists have demonstrated that biochar addition has positive impacts on AD. Biochar addition increases biogas yield and methane concentration and reduces gas contaminants. This influence is attributed to several mechanisms: providing support for microbial colonization, adsorbing excess inhibitory products, and improving electron transfer between syntrophic microorganisms. However, due to the cost and energy-intensive conversion process, the additional costs for biochar addition could offset the resulting energy benefits. This study investigates an integrated biorefinery where corn stover biochar is added to the AD of animal manure to produce biogas-derived Renewable Natural Gas (RNG) for pipeline injection.

A kinetic model of AD, biogas upgrading, and pipeline injection is simulated in Aspen Plus ™ for a mid-size farm with 5000 cattle heads. The process simulation model estimates a 28% increase in biogas yield for 3% (wt. % of wet mass) of biochar addition. High pressurized water scrubbing process is used for biogas upgrading into RNG. Techno-economic analysis (TEA) determines the minimum RNG selling price for a 30-year plant lifetime with a 10% Internal Rate of Return (IRR). TEA estimates the minimum RNG selling price as $42/MMBTU and $35/MMBTU for the without biochar and with biochar scenario, respectively. Biochar addition lowers the RNG selling price by 16% whereas adding revenue from digestate selling and renewable identification number (RIN) credits result in up to 32% reduction. Biochar addition increases the capital and operating costs by 8% and 13%, respectively. Cradle-to-gate life cycle analysis (LCA) is performed in SimaPro ™ to evaluate the GHG emissions of biochar amended AD for RNG production. GHG abatement for the without biochar and with biochar scenario is 0.038 kg $CO_{2\,eq}$/MJ of RNG and 0.123 $CO_{2\,eq}$ /MJ of RNG, respectively. LCA results show biochar addition leads to 2.18 times GHGs abatement compared to the without biochar scenario. Environmental benefits mostly come from the additional carbon sequestration with biochar land use and fossil fuel displacement by syngas from pyrolysis. Sensitivity analysis shows the influence of different major processes and economic parameters on the RNG selling price. The process is cost-prohibitive for small-scale farms due to the significant economy of scale penalty for the biogas upgrading and RNG injection facility. Increased biogas yield reduces the manure input leading to less digestate production. This would decrease the environmental benefits of emissions avoidance from conventional manure management and carbon sequestration per MJ of RNG while reducing the emissions from digestate transportation and applications. Uncertainty analysis considers the variations of key technical and economic parameters within a reasonable range and gives an estimated range of fuel selling prices and GHG abatement. Minimum RNG selling prices vary between $8.82/MMBTU and $16.28/MMBTU and GHG abatement vary between 0.06 kg $CO_{2\,eq}$/MJ and 0.16 kg $CO_{2\,eq}$/MJ within 10-90% confidence intervals.

627985 Optimizing Syngas Production Process By Downdraft Gasifier Using Response Surface Methodology

Abstract:

The gasification process of biomass to produce combustible gases has an attracted interest recently. The downdraft gasifier has an advantage over other types of gasifiers because of its small size, high quality gas production, and low tar yield. One of the challenges in the gasification process is the production of low-quality syngas (low concentration of H2 and CO) and large amount of tar. To overcome this problem, optimization of the gasifier operating conditions was employed to achieve high-quality product gases and hence high gasification process efficiency. In this work, the influence of initial gas temperature, airflow rate, and mole fraction of oxygen in primary air, on the gasification efficiency and product quality was optimized by response surface methodology. The obtained results showed that the reduction in the temperature caused a low concentration of carbon monoxide and a high amount of tar in the product, thereby there was a decline in the efficiency of the gasification process.

623925 Techno-Economic Evaluation of Nano-Porous Silica Production from Rice Husk and Sand

Abstract:

Global rice production in 2019 was estimated at about 500 million tons. Rice husk is a by-product from rice, including about 20 wt% of rice and is needed to be disposed as waste. Rice husk contains approximately 14-25% of ash. The silica content in rice husk ash (RHA) is above 90%. Nanoporous silica (NPS) used in industrial fields such as reinforced rubber additives (tires), concrete mixture (cement binder), nano-fibers, nano-filters, and membranescan be produced from RHA. Rice hisk is combusted at a relatively low temperature (650-800°C) and RHA is produced, whereas the silica sand is operated at 1400-1500°C to produce silicate. In addition, compared to the conventional NPS production from sand, The NPS production from RHA has an advantage in lowering the production cost due to low material cost and energy consumption.

In this study, the techno-economic feasibility for the sodium silicate ($Na_2O{\cdot}xSiO_2$)-based nano-porous silica (SS-NPS) production from sand and rice husk was performed. The return on investment (ROI) and payback period (PBP) were evaluated according to the plant size for the two cases. The effects of raw material and product prices on ROI and PBP were analyzed and economic conditions to be economically viable were suggested.

AIChE
The Global Home of Chemical Engineers

A CAPITAL AND ENERGY EFFICIENT ALTERNATIVE EMISSIONS CONTROL TECHNOLOGY FOR CONTROLLING EMISSIONS USING BEAD ACTIVATED CARBON

Kim K. Tutin, Captis Aire LLC, East Point, GA United States
Harold R. Cowles, Environmental C&C Inc., Scottsdale, AZ United States
John M. Berger P.E., Environmental C&C Inc., Fishers, IN United States

TABLE OF CONTENTS:

I. INTRODUCTION:

In manufacturing, process exhaust gas (air) streams released may contain volatile organic compounds (VOCs), Hazardous Air Pollutants (HAPs), particulate, etc. These are considered pollutants/emissions and must be removed from the air prior to release to avoid air pollution emissions into our communities. Add-on controls either collect the VOCs for recovery and reuse or destroy the VOCs. A number of control options are available including thermal oxidizers, catalytic oxidizers, and biofiltration which destroy pollutants. Other control systems including condensation, scrubbing, and adsorption systems may recover pollutants.

This paper provides an introduction to an adsorption technology, the Fluidized Bed Concentrator (FBC) with Bead Activated Carbon (BAC) media. It describes 1) the FBC technology, 2) the BAC media used in these systems, 3) the performance characteristics, 3) typical applications of this technology, and 4) a case study review of the FBC technology demonstration for controlling emissions and collecting terpenes from wood drying.

II. DESCRIPTION OF FBC TECHNOLOGY: FOUR PRIMARY PARTS

1. Adsorber:

The adsorber section has trays filled with very adsorbent BAC media. Contaminated process exhaust enters from the bottom, passes upward through the adsorption trays, and fluidizes the Bead Activated Carbon (BAC) adsorbent media. When the process air interacts with the BAC, it adsorbs (captures) the terpenes Volatile Organic Compounds (VOCs) in its pores. VOC saturated BAC flows to the bottom of the adsorber vessel and is transferred using fans to pneumatically transfer it to the desorber. Cleaned (reusable) BAC from the desorber feeds continuously into the top of the adsorber, providing counter-current VOC removal. Cleaned air exits out the top of the adsorber.

2. Desorber:

In the desorber, the BAC media is heated (regenerated), causing it to release the VOC vapors into a low volume, gas stream. This gas stream is then circulated continuously to the next stage in the process. The regenerated (cleaned) BAC is then returned continuously to top of the adsorber for reuse. Returning this regenerated BAC to the top of the adsorber provides a final "polishing" of emissions to remove residual emissions prior to release of the process exhaust air.

"All Electric" Option

Desorber Recovery System: In the recovery system, a low volume inert carrier (nitrogen) gas is introduced into the bottom of the desorber. Gas flows up through the desorber as the BAC flows downward. The system is heated via either hot oil or electric heating. The heat causes the VOCs to volatilize off the BAC. The VOC vapors are liberated resulting in a concentrated waste flue gas stream that moves from the desorber to the condenser. This system may eliminate the on-site use of natural gas entirely. Thus, it offers an "All Electric" control technology.

Desorber Oxidation System: In the oxidation system, clean hot stripping air at 100-700°F is introduced to the bottom of the desorber. It flows upward across the BAC media to both volatilize the VOCs and fluidize the BAC. The VOC vapors are liberated resulting in a concentrated waste flue gas stream that moves from the desorber to the oxidizer.

3. Condenser or Oxidizer:

The gas stream from the desorber is circulated continuously to this system for further treatment to either capture and recover VOCs or oxidize them.

"All Electric" Option

Condenser Recovery System: In the recovery system, nitrogen gas circulates from the desorber to the condenser continuously. The VOC vapors are sufficiently concentrated such that the vapors can be cooled via a tube and shell condenser converting the vapor to liquid. The liquid drains continuously into a storage system. The liquid VOCs can then be reused or sold. The nitrogen recirculates back to the desorber where it entrains the VOCs vapors again and the cycle continues.

Oxidizer Oxidation System: In the oxidation system, the vapors are instead sent to a small thermal oxidizer to treat the VOCs. While this system does not provide recovered (collected) VOCs, it nevertheless retains FBC technology benefits including significantly reducing natural gas usage. The burner component of the thermal oxidizer is selected to provide enough energy to increase the thermal oxidizer to the 1400°F operating temperature. The burner is designed to minimize the "fire rate and thus consume negligible fuel. Optionally, a "low NO_x" design can be selected to minimize production of NO and NO_2 (oxides of nitrogen). Thus, the system minimizes energy usage and GHG emissions. [1]

4. Side Stream Reactivator (SSR) [2]:

A small portion of the BAC media exiting the desorber may be transferred to an optional SSR. The SSR reactivates the BAC by heating it to 1450°F with the introduction of a small amount of water which reactivates the BAC continuously to regenerate it and thus maintain its ability to adsorb VOCs. Reactivated BAC from the SSR is transferred back to the top of the adsorber for reuse.

III. ADSORBENT MEDIA:

The BAC adsorbent media utilized in this system is an essential element of the FBC technology. Specialized Bead Activated Carbon (BAC) efficiently adsorbs VOCs (i.e. terpenes) from the process air. The BAC has high-fill (adsorption) capacity, is a small sphere-shaped particle that is less than 1 mm in diameter with a narrow particle size distribution, has low dust, has high purity, has high flowability, and has high strength and wear resistance as measured by ASTM D 3802 – 79 (2005) Standard Test Method for Ball-Pan Hardness of Activated Carbon.[3] The spherical shape enables the BAC to fluidize and flow.

The BAC media can be reactivated and reused. This is accomplished either by either continuous or batch processes.

---

[1] "VOC Removal and Destruction System and Method". U.S. Patent and Trademark Office. (USPTO) https://appft.uspto.gov/netacgi/nph-Parser?Sect1=PTO2&Sect2=HITOFF&p=1&u=%2Fnetahtml%2FPTO%2Fsearch-bool.html&r=1&f=G&l=50&co1=AND&d=PG01&s1=20190022579.PGNR.&OS=DN/20190022579&RS=DN/20190022579 (Accessed October 4, 2021).

[2] "GASTAK® Solvent Recovery and Odor Removal by Fluidized Bed Process". Kureha Ecology Management Co. , LTD website. https://www.kurekan.co.jp/en/download/gastak.pdf (Accessed September 18, 2021).

[3] "ASTM D3802-79 (2005) Standard Test Method For Ball-Pan Hardness Of Activated Carbon. https://webstore.ansi.org/standards/astm/astmd3802792005 (Accessed September 18, 2021).

Continuous Reactivation: In the continuous process, a small amount of BAC media is diverted from the BAC exiting the desorber and sent to an in-line side stream reactivator (SSR).[4] It is then returned to the top of adsorber. Returning this reactivated BAC to the top of the adsorber provides a final "polishing" of emissions to remove residual emissions prior to release of the process air.

Batch Reactivation: In the batch process, for operators that do not have an SSR, the BAC media is unloaded periodically, shipped offsite for reactivation, and then reloaded once returned. In the interim, a second load of BAC is utilized. Reactivation is typically performed by California Carbon who has been providing reactivation services since 1975.[5] This service includes a reactivation warranty *"Carbon will be reactivated to 95% of Virgin Activity per ASTM D 3467 and Carbon will lose no more than 10% of incoming volume of the specified mesh size in the reactivation process."*[6] BAC lifetime is site and application specific. Some operators have been reusing their BAC for years.

Bead Activated Carbon (BAC) Adsorbent Media

Bead Activated Carbon (BAC), the Adsorbent Media, Fluidized in Tray

IV. FBC OPTIONS: RECOVERY AND OXIDATION

The Fluidized Bed Concentrator can be utilized to either recover the VOCs or oxidize them as shown in the figures. Operators typically select the desired option based on a number of factors especially including the anticipated value and reusability of the recovered VOCs. The value of the VOCs depends on a number of factors including the anticipated quantity recovered, purity, salability, and/or reusability.

Recovery: When the value of the recovered VOCS is high and if the cost of recovery is less than the cost of purchasing new material, then recovery makes sense.

[4] "GASTAK® Solvent Recovery and Odor Removal by Fluidized Bed Process". Kureha Ecology Management Co. , LTD website. https://www.kurekan.co.jp/en/download/gastak.pdf (Accessed September 18, 2021).
[5] "California Carbon Co. Inc. Welcome". California Carbon website. http://www.californiacarbon.com/index.html (Accessed September 18, 2021).
[6] "Custom Spent Carbon Reactivation". California Carbon website. http://www.californiacarbon.com/pdfs/Custom%20Reactivation.pdf (Accessed September 18, 2021).

Oxidation: If the VOC has low or no recovery value, as, for example, when it is a mixture, or if it poses disposal concerns, such as for toxic compounds, then destruction probably makes the most sense. Excluding value derived from VOC recovery, the oxidation option retains FBC system performance characteristics including reduction in energy usage.

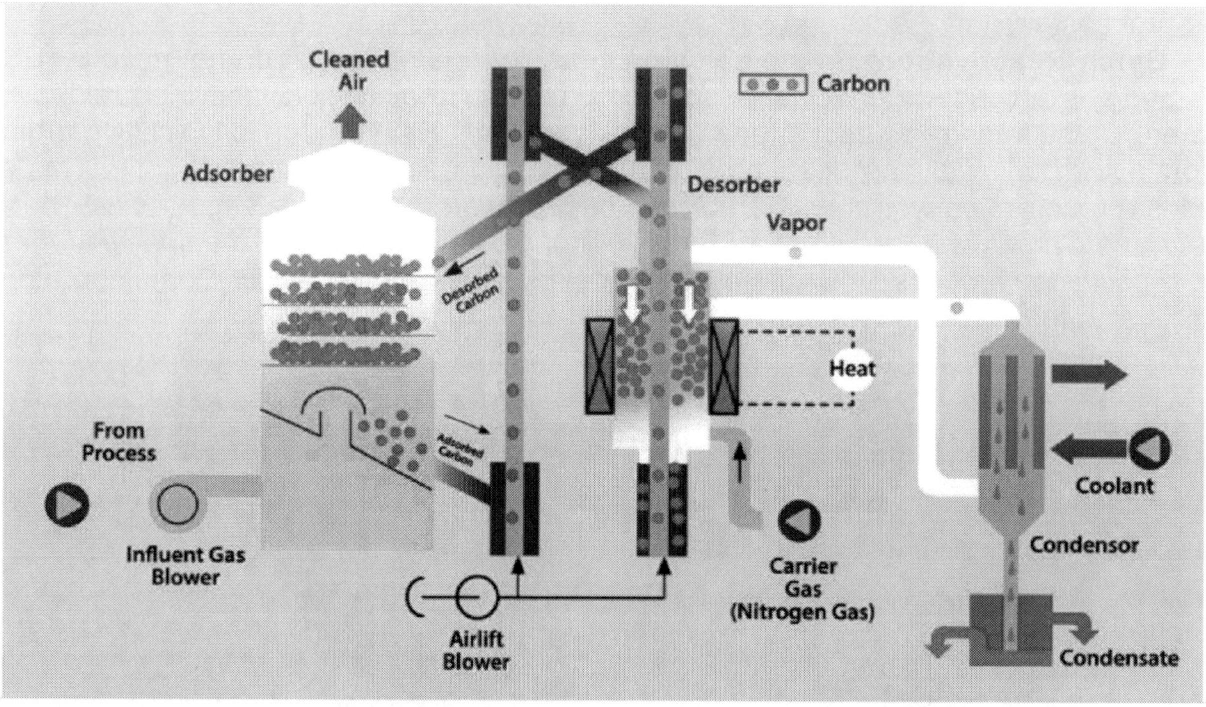

RECOVERY SYSTEM: Adsorbs, Desorbs, and Condenses VOCs

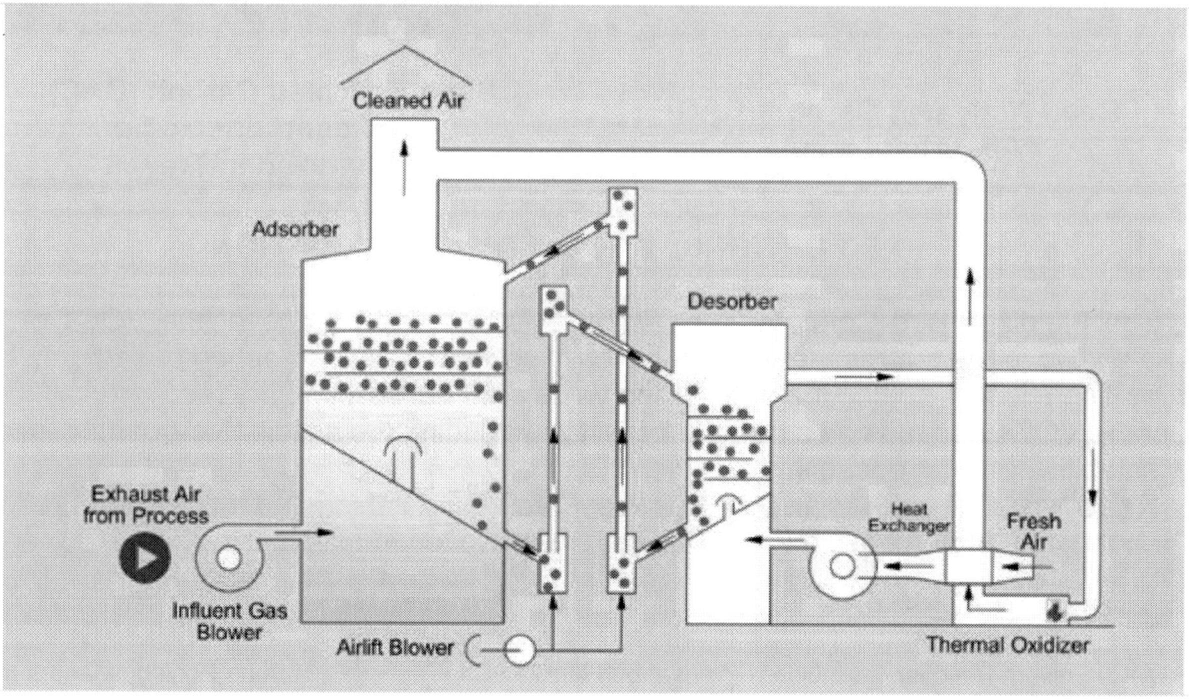

OXIDATION SYSTEM: Adsorbs, Desorbs, and Oxidizes VOCs

V. PROCESS PERFORMANCE CHARACTERISTICS:

1. The FBC reduces energy usage (i.e. natural gas) versus oxidation systems due to reduction or elimination of energy used to heat the air going through thermal oxidizers:

Baseline Oxidation Systems:

Thermal oxidizers burn VOCs by heating them in enclosed combustion chambers in the presence of excess oxygen. The precise temperature and residence time needed for effective reduction of VOC emissions varies with the nature of the VOC. However, temperatures of 1400-2000°F and residence times of 0.5-2 seconds are typical. Since auxiliary fuel is necessary to maintain combustion temperatures, most thermal oxidizers incorporate some form of heat recovery to lower auxiliary fuel costs.

FBC Technology:

The air going through the adsorber is not heated. Thus, the natural gas usage is reduced or eliminated.

Recovery – "All Electric" Option: For a recovery system, the control device can be "All Electric". It could eliminate the use of natural gas entirely.

Oxidation: The FBC oxidation system significantly reduces natural gas usage but does not eliminate it.

Energy Reduction Calculation: The amount of energy that is eliminated, that would have been used by the thermal oxidizer, can be calculated utilizing one of many credible sources. One of the best sources is the Institute of Clean Air Companies' Guidance Methodology[7]. Calculators are also available on-line including those provided by the EPA[8], Cycle Therm[9], and Captis Aire[10]. With these calculators, users can input their specific data and the calculators will calculate the estimated amount of energy required to run the thermal oxidizer. All these sources rely on a single primary factor to calculate energy required to heat the air. This factor is the specific heat of air which is the number of British Thermal Units (BTUs) required to heat one pound of air by one °F. This factor along with the system specific factors including the standard cubic feet per minute (sCFM) of air, the temperature increase, the density of air, and conversion factors provide the number of BTUs required to heat the air going through the device. These factors are used together with the energy efficiency and the heat generated by the oxidation of the emissions to calculate the BTUs of energy required to heat the air going through the RTO. Users can then multiply BTUs times the forecasted long-term cost of natural gas to estimate the long-term cost of natural gas required over the 20+ year life of the thermal oxidizer. This calculation provides the value of the natural gas elimination.

---

[7] "Institute of Clean Air Companies (ICAC) Guidance Method for Estimation of Gas Consumption in a RTO". Institute of Clean Air Companies website. https://cdn.ymaws.com/www.icac.com/resource/resmgr/RTO-F1.pdf (Accessed September 18, 2021)

[8] "Air Pollution Control Cost Estimation Spreadsheet". Environmental Protection Agency (EPA) website. https://www.epa.gov/sites/production/files/2018-01/oxidizers_calc_sheet_finalversion_1-16-2018.xlsm (Accessed Feb 3, 2021).

[9] "Operating Cost Calculator". Cycle Therm website. http://www.cycletherm.com/resources/operating-cost-calculator.aspx (Accessed Feb 3, 2021).

[10] "Calculate Engineering Estimate of Natural Gas Usage and Greenhouse Gas Generation". Captis Aire website. https://www.captisaire.com/Value-Calculator/ (Accessed Feb 3, 2021).

2. The FBC generates revenue from sales of recovered organics:

When using a recovery system, including the FBC, operators have the option to collect and sell or reuse the volatile organic compounds recovered. The liquid VOCs are collected from the FBC condenser for reuse or sale. The value of recovered VOCs will depend on the type, amount, and purity of VOCs collected.

3. The FBC reduces Greenhouse Gases (GHG) emissions versus oxidation systems due to the reduction or elimination of burning natural gas:

Due to the reduction or elimination of natural gas usage and reduction in electrical usage by the FBC, greenhouse gas emissions are also reduced. The greenhouse gas emissions reduction can be calculated based on the amount of carbon in natural gas eliminated and its stoichiometric conversion to carbon dioxide (CO_2). The Environmental Protection Agency (EPA) has an on-line calculator[11] that enables the user to calculate greenhouse gas emissions that are eliminated by eliminating natural gas used to heat the air in the thermal oxidizer.

4. The FBC enables creation of carbon offsets versus oxidation systems:

In conversations with American Carbon Registry[12] and ClimeCo[13], it may be possible to measure, register, and verify carbon offsets for a site converting to the FBC technology from an oxidation system due to the reduction in energy usage.[14]

The quantification methodology and value of carbon offsets is evolving quickly. But these carbon offsets can be sold or traded externally. Or they may be utilized for internal credit accounting purposes. Potential values for these carbon offsets vary widely.[15] However, as of early 2021, American Carbon Registry and ClimeCo estimated the value of carbon offsets for an FBC project at $5/ton.[16]

5. The FBC reduces electricity usage versus oxidation systems:

The horsepower required to run the fans to move the process exhaust air through the BAC media is reduced for the FBC due to the fact that it's BAC is fluidized (moving) rather than packed (non-moving). The pressure drop across the system is reduced from a typical 10-20" to 3.5". Thus, the horsepower required to run the fans for the FBC may be ~ ¼ of that required for the RTO.

To estimate the electrical savings, the sum of electrical demands from the FBC is subtracted from the sum of electrical demands of the thermal oxidizer. Net kW·h along with site-specific electricity costs and operating hours per year are used in combination to estimate electrical savings.

---

[11] "Greenhouse Gases Equivalencies Calculator - Calculations and References". EPA website. https://www.epa.gov/energy/greenhouse-gases-equivalencies-calculator-calculations-and-references (Accessed December 28, 2020).

[12] "Harnessing the Power of the Markets to Improve the Environment". American Carbon Registry website. https://americancarbonregistry.org/ (Accessed February 3, 2021).

[13] "Scalable greenhouse gas projects to earn real financial returns for your business." ClimeCo website. https://climeco.com/ (Accessed February 3, 2021).

[14] "AM0044: Energy efficiency improvement projects - boiler rehabilitation or replacement in industrial and district heating sectors --- Version 2.0.0". Clean Development Mechanism (CDM) website. https://cdm.unfccc.int/methodologies/DB/3HZ4USHZ2W449HMAXZN420E5PJB1QF (Accessed Sept 18, 2021)

[15] "Allowance Price Explorer". International Carbon Action Partnership website. https://icapcarbonaction.com/en/ets-prices (Accessed Jan 20, 2021).

[16] Eric Ripley (Director Industrial Programs, American Carbon Registry), in a phone conversation with Kim Tutin November 17, 2020.

6. **<u>The FBC enables superior mass transfer of VOCs through counter current flow fluidized BAC media that is continuously regenerated:</u>**

Baseline Adsorption Systems:

Adsorption: In sorbent treatment systems, process exhaust air moves across the sorbent. VOCs are removed via attraction (adsorption) of the VOCs on carbon or zeolite sorbent media in packed (non-moving) beds. This may be accomplished by concentrators that have wheels with sectors containing activated carbon or zeolite media. In this arrangement, most of the space on the wheel functions in adsorption mode, with one small sector being regenerated at any given time. Or the process exhaust air may be forced through large vessels containing sorbent media. In this arrangement, the VOCs are treated via a once-through process.

Regeneration / Reactivation: Regeneration is accomplished by heating the sorbent media with heaters, heated gases, steam, etc. If using steam, water may be a byproduct. This can be accomplished either via off-site regeneration or by alternating the flow between multiple vessels where one vessel functions in adsorption mode while the other is being regenerated at any given time. In both these arrangements, the sorbent media is being regenerated in batch or semi-continuous fashion. Reactivation may be accomplished by sending the sorbent media off site for reactivation. Or the sorbent media may be discarded when it becomes saturated.

Effect of water vapor: Especially for process exhaust air heavily laden with water vapor, i.e. 10% B_{ws} moisture by volume[17] or more, the sorbent media may become partially or fully saturated with water. At high moisture content, water molecules begin to compete with the hydrocarbon molecules for active adsorption sites. This reduces the capacity and the efficiency of the adsorption system.

FBC Technology:

Adsorption: The BAC media in the fluidized bed concentrator adsorber is fluidized (moving) by the process air moving opposite (counter current) to the BAC as it moves downward through the trays in the adsorber. This provides excellent mixing of the BAC and contact with the air. As this occurs, VOCs or odors are transferred from the process air to the BAC. The cleaned air exits through the top of the adsorber.

Regeneration / Reactivation: The BAC media is regenerated continuously via continuous recirculation to and from the desorber. It is reactivated continuously if system includes the Side Stream Reactivator (SSR). This ensures the BAC retains the capacity to maintain adsorption efficiency. This is done at temperatures well above the boiling point of water, often around 400°F or more. This minimizes or eliminates accumulation of water and/or organics on the BAC.

Effect of water vapor: Continuous regeneration and reactivation at temperatures above the boiling point of water, i.e. 400°F, volatilizes the water off the BAC media and prevents the BAC from becoming saturated with water. Thus, it maintains adsorption efficiency even when treating process exhaust streams with high moisture contents, as are typical for wood drying applications, i.e. ≥20% B_{ws} moisture by volume[18]. This system minimizes or eliminates need for dehumidification systems.

---

[17] "METHOD 4—DETERMINATION OF MOISTURE CONTENT IN STACK GASES". EPA website. https://www.epa.gov/sites/default/files/2016-06/documents/m-04.pdf. (Accessed October 17, 2021).
[18] "METHOD 4—DETERMINATION OF MOISTURE CONTENT IN STACK GASES". EPA website. https://www.epa.gov/sites/default/files/2016-06/documents/m-04.pdf. (Accessed October 17, 2021).

Effect of Fluidized (moving) vs Packed (non-moving) media: BAC media fluidization prevents pockets of high concentration VOCs, provides higher dry particulate tolerance (reduced plugging of interstitial spaces between media), and enables continuous on-site regeneration and reactivation.

TABLE I
PHYSICAL AND OPERATIONAL COMPARISON OF THE FBC VS THE RTO

Parameters:	RTO	FBC	Advantages of FBC
Treatment Temperature, °F	>1400°F	175°F	• Adsorption at process air temperature • No natural gas required for heating
Media Desorption Temperature, °F	>1400°F	400°F	• Much less energy required for desorption • Continuous media regeneration, above the boiling point of water, prevents the media from getting saturated with water. Thus, it maintains its ability to adsorb pollutants.
Media depth, feet (ft)	16 ft	1 ft	• Less Electricity (fan HP) required to move air through less media
Media Configuration	Packed (Not Moving)	Fluidized (Moving)	• Less Electricity (fan HP) required to move process air through fluidized media
Pressure Drop, Inches (")	10-20"	3.5"	• Less Electricity required to move process air against less pressure
Fan size, Horsepower, (HP)	2400 HP	600 HP	• Less HP due to media that is fluidized. • Less susceptible to plugging from particles
Emissions Treatment	Oxidation to CO^2	Adsorption (Capture)	• Emissions are NOT oxidized to CO_2, so reduces GHG emissions • Emissions collected can be sold or reused
Media Reuse	Discarded	Reused	• Media desorption with 'in unit' continuous regeneration. • Flowable bead shaped media can be removed and reused
Safety		Inherently Safer	• Lower operating temperatures • No large heavy moving parts • Easier cleaning and maintenance

VI. AWARDS:

The FBC has been recognized for excellence in environmental performance. Honda Motors of Alabama was recognized as *"Air Conservationist of the Year"* when they implemented the FBC to control emissions when they increased production.[19] Ford's "Fumes to Fuels" work was awarded the EPA's *"Clean Air Excellence"* award.[20] In this work, they utilized the FBC to capture solvent-laden air, strip the solvent from the air, concentrate the solvent up to 2,000 times, and send it to a fuel reformer. The fuel reformer used the concentrate to generate hydrogen-rich fuel, which was directed into a fuel cell to generate electricity.

VII. APPLICATIONS:

Concentration systems, including the FBC, are advantageously used in front of control devices when treating very high volumes of very dilute exhaust gas to reduce energy usage. Recovery systems are beneficial when the value of the recovered VOCs is high. The following are a few examples where the FBC technology has been utilized historically.

TABLE II
FBC APPLICATIONS

Application	Chemistry
Adhesive Tape Mfg	Toluene
Agricultural Products	Organic Acid & Ammonia Odor
Aluminum Casting	Phenol, Formaldehyde, Ammonia
Animal Lab	Odor Control
Atomic Power Plant	Styrene
Auto Parts Degreasing	Trichloroethylene
Bakery	Odor Control
Brewery	Odor Control
Car Wash	Kerosene
Cellophane Coating	Toluene, Ethyl Acetate, Butyl Acetate
Ceramic Condenser	Terpineol
Confectionery	Odor control
Chemical Production	Toluene, Higher Molecular Weight Alcohols, Carbon Tetrachloride

[19] "HMA Honored as Air Conservationist of the Year". Honda website. https://www.hondaalabama.com/article/hma-honored-as-air-conservationist-of-the-year (Accessed September 21, 2021).
[20] "Clean Air Excellence Award Recipients". Environmental Protection Agency (EPA) website. https://www.epa.gov/sites/default/files/2015-06/documents/clean_air_excellence_award_recipients_year_2003.pdf (Accessed September 21, 2021).

TABLE II (CONTINUED)
FBC APPLICATIONS

Application	Chemistry
Dye Production	Perchloroethylene
Electronics Mfg Plant	DMF, MEK, etc.
Electronics Parts Cleaning	Trichloroethylene, Toluene, 1,1,1-Trichloroethane
Fabric Washing	Perchloroethylene
Feed Processing	Odor Control
Film Coating	MEK, Methyl Cellosolve, Toluene, Tetrahydrofuran
Film Laminating	Toluene, Ethyl Acetate, n-Hexane
Gravure Printing	Toluene, Xylene, n-Propanol, n-Propyl Acetate, mixed
Iron Casting	Phenol, Formaldehyde, Ammonia
Lacquering of Film	Toluene, Ethyl Acetate, MEK
Landfill	Odor Control
LSI Mfg Plant	Phenol, Acetone, Methanol, Mixed Solvents
Lubrication Mfg	Mixed Solvent Odor Control Magnetic Tape, Chlorinated Hydrocarbon
Medicine Mfg Plant	Butanol, Methylene Chloride
Paint Booth	Thinner
Printing Ink Mfg	Mixed Solvents
PVC Resin Mfg Plant	Vinyl Chloride
Resin Mfg	N-Hexane, Methylene Chloride
Rubber Coating	Toluene
Rubber Vulcanizing	Odor Control
Sandpaper Mfg	Toluene, Xylene, Ethyl Acetate
Semiconductor Mfg	IPA, HMDS, NMP, Ethanol, Methanol, Ethyl lactate
Silk Screen Printing	Xylene, Mixed Solvents
Teflon Tubing Mfg	Perchloroethylene
Wastewater Aeration	Odor Control

TABLE III
EXAMPLES OF FBC SYSTEMS INSTALLED

FLUIDIZED BED CONCENTRATOR SYSTEMS INSTALLED: [21]	
Industry **Location** **Inlet Stream** **VOCs** **Performance Achieved** **Notes**	Semiconductor Chip Manufacturing Israel 20,000 CFM Semiconductor solvents, 50 ppmv 95% Reduction Efficiency System located in populated area of a major city has eliminated nuisance odor
Industry **Location** **Inlet Stream** **VOCs** **Performance Achieved** **Notes**	Semiconductor Chip Manufacturing Arizona 36,000 CFM Semiconductor solvents, 25-100 ppmv 90%+ Reduction Efficiency System met strict local compliance requirements while achieving Zero Unscheduled Downtime (ZUD)
Industry **Location** **Inlet Stream** **VOCs** **Performance Achieved** **Notes**	Teflon Tubing Manufacturing Massachusetts 1,800 CFM Perchloroethylene, 1500 ppmv 98% Reduction Efficiency Recovered solvent re-used in process. System payback in 1 year.
Industry **Location** **Inlet Stream** **VOCs** **Performance Achieved** **Notes**	Furniture Painting California 50,000 CFM Paint Solvents, 25-50 ppmv 90%+ Reduction Efficiency Recovered solvent and reformulated with paint
Industry **Location** **Inlet Stream** **VOCs** **Performance Achieved** **Notes**	Military Vehicles Michigan 150,000 CFM Paint Solvents 95% Reduction Efficiency Controlled multiple different interchangeable VOC sources.

[21] "Environmental C&C Inc". Environmental C&C website. https://www.environmentalcc.com/ (Accessed September 18, 2021).

Automotive Site, Alabama, 100,000 CFM
Paint Solvents, >90% Reduction Efficiency
Two adsorbers and one desorber

Image shows three adsorbers one desorber	
Image shows ductwork that enabled control of any of the 3 sources to any of the 3 adsorbers. **Each adsorber has its own process fan.** **System provided control of multiple different interchangeable VOC sources**	

Military Vehicles, Michigan, 150,000 CFM
Paint Solvents, 95% Reduction Efficiency

TABLE IV: VALUE CREATION CALCULATION EXAMPLE

Fluidized Bed Concentrator (FBC) vs Regenerative Thermal Oxidizer (RTO)	Value per Year
1) Reduce Energy Usage (Natural Gas) due to Elimination of Energy Used to Heat the Air in the RTO: [22] [23] a. $\frac{mmBTU}{Year}\ to\ heat\ air\ in\ RTO = (\Delta\ \mathrm{T})\ (sCFM)(1.08)(\%Heat\ Loss)(\frac{Operating\ Hours}{Year})/1000000$ b. $(1500°F - 175°F) * (300{,}000sCFM) * \left(\frac{1.08\ BTUs}{sCFM \cdot Hour \cdot Degree°F}\right) * (7.5\%) * (8000\frac{Hrs}{Yr})/1000000$ c. $= \left(257{,}580\frac{mmBTU}{Yr}\right) - [\left(250\frac{lbs\ VOC}{Hour}\right) * \left(15{,}000\frac{BTUs}{lb\ VOCs}\right) * \left(\frac{8000\ Hrs}{Year}\right)] = \mathbf{227,580}\frac{\mathbf{mmBTUs}}{\mathbf{Year}}$ d. $\left(\frac{227{,}580\ mmBTUs}{Year}\right)(\$3.5\frac{Long\ Term\ Cost\ of\ Natural\ Gas}{mmBTU}) = \mathbf{\$796,530\ Savings/Year}$	Save by Eliminating Natural Gas **$796,530**
2) Reduce Greenhouse Gases (GHG) Generated from Burning Natural Gas: [24] a. $Tons\ GHG = (mmBTU) * \left(\frac{117\ lbs\ CO2}{mmBTU}\right) * (\frac{1\ Metric\ Ton}{2205\ lbs})$ b. $\left(227{,}580\frac{mmBTUs}{Yr}\right) * \left(117\ lbs\frac{CO2}{mmBTU}\right) * \left(1\frac{Metric\ Ton}{2205\ lbs}\right) = \mathbf{12,076}\frac{\mathbf{Tons\ GHG\ Reduction}}{\mathbf{Year}}$	Sell Carbon Offsets **$60,378**
3) Sell Carbon Offsets (Carbon Credits) Utilizing Carbon Accounting Service Providers, i.e. ClimeCo: [25] a. $\left(\frac{12{,}076\ Tons\ GHG}{Year}\right) * \left(\frac{\$5}{Ton\ GHG}\right) = \mathbf{\$60,378/Year\ Selling\ Carbon\ Offsets}$	
4) Reduce Electricity Usage by Reducing Fan Power to Move Process Air through Fluidized Media: $(2400RTO - 600FBC\ HP\ fans - 800\ HP) * \left(\frac{0.75\ kW \cdot h}{HP}\right)\left(\frac{\$0.07}{kW \cdot h}\right)\left(\frac{8000\ Hrs}{Year}\right) = \mathbf{\$420,000}\frac{\mathbf{Savings}}{\mathbf{Year}}$	Save by Reducing Electrical Usage **$420,000**
5) Sell Terpenes [26] **Yielded from Green Pine Wood** [27]**:** a. $\left(\frac{1\ Gallon\ Terpenes}{Oven\ Dry\ Ton\ Wood}\right) * \left(\frac{300{,}0000\ Dry\ Tons\ Wood}{Year}\right) * (75\%Yield) * \left(\frac{\$3}{Gal}\right) = \mathbf{\$675,000/Yr}$ **Selling Terpenes**	Sell Terpenes **$675,000**
6) Reduce Maintenance Costs:[28] Media changeouts, large heavy moving part replacement, corrosion repairs, etc.	Save up to **~$1 Million**
Total Value Creation per Year per Site:	**Up to $3 Million**

---

22 ICAC Guidance Method for Estimation of Gas Consumption in a RTO. Institute of Clean Air Companies website. https://cdn.ymaws.com/www.icac.com/resource/resmgr/RTO-F1.pdf (Accessed September 18, 2021)

23 "Operating Cost Calculator". Cycle Therm website. http://www.cycletherm.com/resources/operating-cost-calculator.aspx (Accessed November 21, 2021).

24 "Greenhouse Gases Equivalencies Calculator - Calculations and References". Environmental Protection Agency (EPA) website. https://www.epa.gov/energy/greenhouse-gases-equivalencies-calculator-calculations-and-references (Accessed December 28, 2020).

25 "Allowance Price Explorer". International Carbon Action Partnership website. https://icapcarbonaction.com/en/ets-prices (Accessed Jan 20, 2021).

26 "Price Chart for Gum Turpentine". PineChem website. http://www.pinechem.net/images/gall/ChartGNE.GIF Accessed Jan 3, 2021).

27 "Sulfate Turpentine Recovery". Pulp Chemicals Association. Pages 9-12, 127. https://www.pinechemicals.org/store/viewproduct.aspx?id=1579188 (Accessed Jan 2, 2021).

28 "Air Pollution Control Technology Fact Sheet". EPA website. https://www3.epa.gov/ttncatc1/dir1/fregen.pdf (Accessed October 17, 2021).

VIII. CASE STUDY: FBC TECHNOLOGY SPECIFIC TO WOOD DRYING

DEMONSTRATION DESCRIPTION:

Captis Aire LLC founder and Environmental C&C Inc. worked on a project together from 2015-2021 with a commercial oriented strand board (OSB) manufacturer to demonstrate the ability to utilize the FBC to control emissions from wood drying.

In wood products manufacturing, process gas exhaust streams released from drying wood contain VOCs, including turpentine (terpenes). These VOCs are considered pollutants/emissions and must be removed from the air, prior to release, to avoid air pollution emissions into our communities. Wood drying operations utilized in the manufacture of oriented strand board (OSB), lumber, wood to biofuels, i.e. wood pellets, plywood, particleboard, etc. are few examples of these processes that have process exhaust streams containing VOCs.

Work on this patent[29] pending[30] technology was executed in three phases. In the first phase, laboratory screening work was done to explore whether emissions from wood drying, primarily terpenes, could be adsorbed by Bead Activated Carbon (BAC) media at the expected operating temperatures. In the second phase, a small-scale pilot adsorber was installed at a commercial Oriented Strand Board (OSB) manufacturing site to demonstrate ability to adsorb emissions from wood drying under commercial manufacturing conditions. In the third phase, the pilot unit capabilities were expanded to include the adsorber, desorber, and condenser. The SSR performance was simulated by semi-continuous additions of virgin BAC during air emissions compliance testing.

In the third testing phase, the pilot unit ran for 6 months with over 1700 hours on-line. Ability to meet air emissions control compliance, per the Plywood and Composite Wood Products Maximum Achievable Control Technology (PCWP MACT) standard, was determined by two independent certified air emissions[31] stack testing companies[32] during two separate testing events.

---

[29] "METHODS AND SYSTEMS FOR RECOVERING TERPENE COMPOSITIONS FROM WOOD DRYING EXHAUST". World International Patent Office (WIPO) Website. https://patentscope.wipo.int/search/en/detail.jsf?docId=WO2019148208 (Accessed September 18, 2021).

[30] "METHODS AND SYSTEMS FOR CONTROLLING EMISSIONS FROM WOOD DRYING PROCESSES". WIPO Website. https://patentscope.wipo.int/search/en/detail.jsf?docId=WO2019148207 (Accessed September 18, 2021).

[31] "AIR Advanced Industrial Resources". AIR website. https://airtest1.com/ (Accessed April 16, 2020).

[32] "Testing and Modeling". John Zink Hamworthy Combustion website. https://www.johnzinkhamworthy.com/testing-modeling/ (Accessed Feb 3, 2021).

TABLE V
RESULTS OF AIR POLLUTION CONTROL COMPLIANCE TESTING (FBC UNIT):

Emissions Measured:	FBC %Reduction Efficiency[33]
Total Hydrocarbons (THC), as carbon including Methane	92%
Volatile Organic Compounds (VOCs), as carbon. (Total non-methane hydrocarbons – THC minus Methane)	92%
Formaldehyde	83%
Methanol	75%
GHC - Nitrogen Oxides (NOx)	55%
Particulate Matter (PM), total filterable	50%
Condensable Particulate Matter	59%
Total Particulate Matter, Filterable + Condensable	51%
Total PM2.5, Particle Size Analyses (PSA) Filterable + Condensable	60%
Total PM10, Particle Size Analyses (PSA) Filterable + Condensable	64%
Carbon Monoxide (CO)	7%
Sulfur Dioxide (SO2)	99%

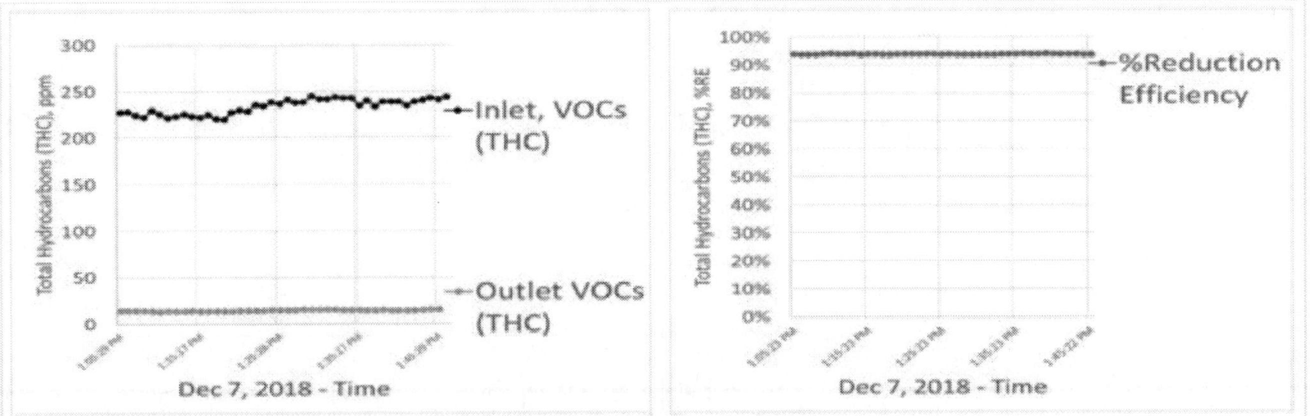

RESULTS OF VOC REDUCTION EFFICIENCY PERFORMANCE MEASURED BY FLAME IONIZATION DETECTOR (FBC DEMONSTRATION UNIT):
Achieved >90% Reduction Efficiency meeting PCWP MACT Standard[34]

33 Emissions results FBC 2018 at commercial wood drying site done by Advanced Industrial Resources, Inc. Results were demonstrated during three 1-hour testing intervals. https://airtest1.com/ . (Accessed September 18, 2021)

34 "Electronic Code of Federal Regulations (eCFR)". eCFR website. See https://www.ecfr.gov/cgi-bin/text-idx?node=sp40.13.63.dddd#ap40.14.63_12292.1 (Accessed April 16, 2020).

RESULTS OF VOLATILE ORGANIC COMPOUNDS CAPTURE:

Terpenes were collected from the FBC demonstration unit. These terpenes are a mixture of terpene hydrocarbons, primarily Alpha and Beta pinenes. These terpenes are responsibly-recovered, biobased materials. Potential applications for these terpenes include use in fuels, fragrances, solvents, adjuvants, cleaners, and polyterpene resins.

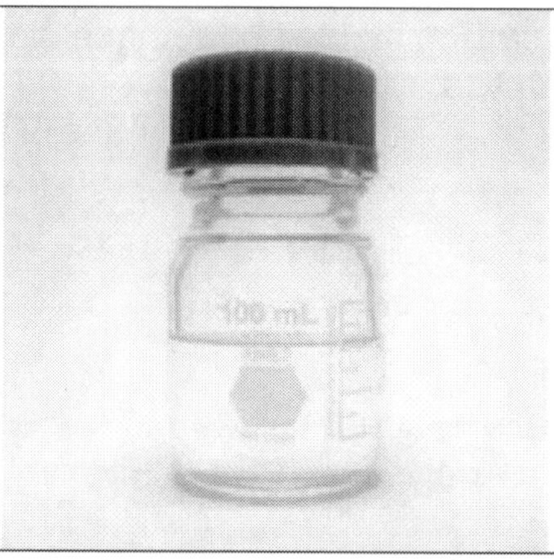

TERPENES COLLECTED FROM FBC DEMONSTRATION UNIT

VALUE OF TERPENES:

The Pine Chemicals Association has documented the gallons of terpenes per oven dry ton of wood in a variety of tree species. Typically, there is ~1 gallon of turpentine (terpenes)[35] per oven dry ton of wood product produced from green pine wood from the SE part of the United States, i.e. Loblolly Pine, Jack Pine, Slash Pine, Shortleaf Pine, Ponderosa Pine, Pitch Pine, Pond Pine, Red Pine, and Spruce Pine. If 75% of this is evaporated and captured from the drying process, the value of the terpenes per year can be calculated by multiplying this yield by the oven dry tons (ODT) of wood produced and a typical price per gallon[36] for terpenes. For example:

$$\left(\frac{1\ Gallon\ Terpenes}{ODT\ Wood}\right) * \left(\frac{ODT\ Wood}{Year}\right) * (\%Yield) * \left(\frac{\$\ Dollars}{Gallon\ Terpenes}\right) = \$\ /\ Year$$

TABLE VI
TERPENES TYPICAL PROPERTIES:

Property	Result
Appearance	Clear Liquid
Moisture	<0.1%
Weight per Gallon	7.2 pounds per gallon
Specific Gravity	0.86
Sulfur Content	< 10 ppm

[35] "Sulfate Turpentine Recovery". Pulp Chemicals Association. Pages 9-12, 127. https://www.pinechemicals.org/store/viewproduct.aspx?id=1579188 (Accessed Jan 2, 2021).

[36] "Price Chart for Gum Turpentine". PineChem website. http://www.pinechem.net/images/gall/ChartGNE.GIF Accessed Jan 3, 2021).

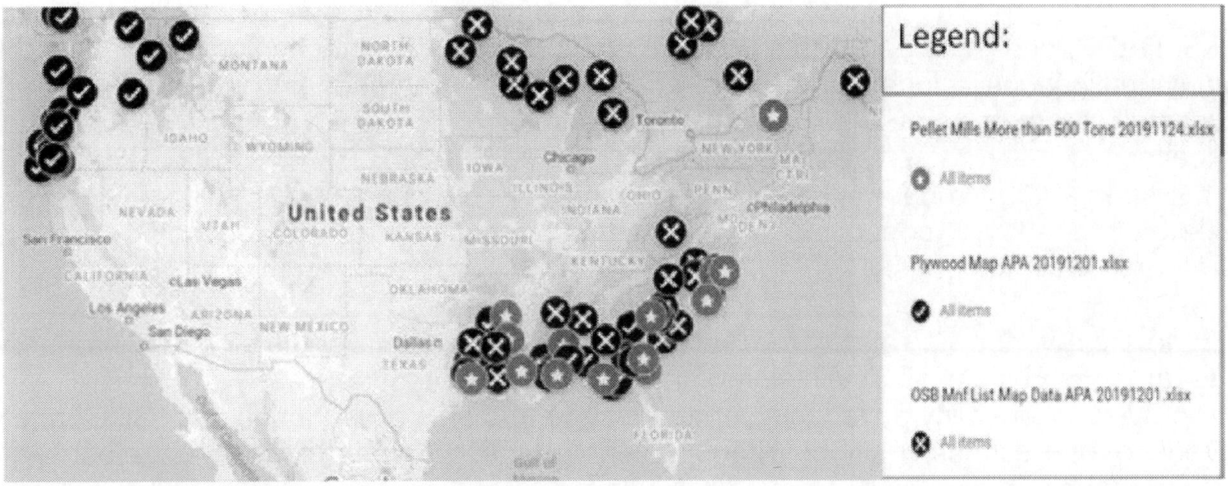

MAP OF POTENTIAL WOOD DRYING SITES FOR FBC USE:

This map shows potential sites where various wood products are produced where the FBC may be used to control emissions. Potential sites include but are not limited to Oriented Strand Board (OSB), lumber (sawmills)[37], plywood[38], wood to biofuels, i.e. wood pellets[39], particleboard, Medium Density Fiberboard (MDF), etc.

RESULTS OF DISCUSSIONS WITH ENVIRONMENTAL AGENCIES:

Official stack testing results were discussed with two state environmental agencies and two federal environmental protection agency (EPA) personnel. In all discussions, environmental agency personnel were supportive of providing a permit for full size commercial FBC unit.

FINAL RESULTS FROM WOOD DRYING DEMONSTRATION. THE UNIT:

- Provided salable terpenes, composed primarily of Alpha and Beta Pinenes.
- Reduced greenhouse gas (GHG) emissions.
- Reduced energy usage. It eliminated natural gas usage and reduced electrical usage.
- Met the PCWP MACT requirement of >90% Reduction Efficiency[40] in both tests. In the second test, the system met the compliance requirements three times in a row, over each of the three one-hour testing intervals.
- Demonstrated good operational reliability.
- Enabled State and Federal environmental personnel that reviewed these results to support permitting for a full commercial FBC system.

---

[37] "The mill map allows you to visually locate and drill down to find the mill you're interested in." Primary Forest Products Network website. https://primary.forestproductslocator.org/mill-map (Accessed September 18, 2021).
[38] "Manufacturer Directory". Engineered Wood Association website. https://www.apawood.org/manufacturer-directory (Accessed September 18, 2021).
[39] "Biomass Magazine". Biomass Magazine website. See http://biomassmagazine.com/plants/map/pellet/ (Accessed September 18, 2021).
[40] "Electronic Code of Federal Regulations (eCFR)". eCFR website. See https://www.ecfr.gov/cgi-bin/text-idx?node=sp40.13.63.dddd#ap40.14.63_12292.1 (Accessed April 16, 2020).

IX. CONCLUSIONS:

The FBC provides an affordable, reliable, and accessible clean technology for controlling VOCS, HAPS, and even particulate in some applications. This technology is well established in many applications including automotive and electronics manufacturing where it has been used successfully for many years. It is an emerging technology for use in biomass (i.e. wood drying) applications.

Especially for process exhaust streams with low VOC concentrations, the FBC technology is a patent[41] pending[42], cost competitive, industrial air pollution control technology that enables operators to advantageously:

1. Generate revenue via sales of recovered organic products (including biobased products such as terpenes in biomass applications) and sales of carbon offsets
2. Improve sustainability by reducing energy usage and thus greenhouse gas (GHG) emissions
3. Reduce operational costs by reducing energy usage (natural gas and electricity) and maintenance costs.
4. Enhance VOC mass transfer to the BAC media
5. Provide continuous in-situ BAC media regeneration and reactivation. This enables the media to continue to be reused.
6. Minimize water adsorbed on the BAC media thus maintaining its ability to adsorb VOCs, even in high moisture content process exhaust streams

Like any air pollution control technology installation, careful consideration of many design parameters, the unit's specific operating profile, and the operator's goals need to be taken into account during the design, installation, and startup/optimization processes. There are excellent resources, including Environmental C&C Inc.[43], Captis Aire LLC[44], and the Institute of Clean Air Companies (ICAC)[45] personnel, to assist with evaluating, implementing, and optimizing FBC applications.

---

[41] "METHODS AND SYSTEMS FOR RECOVERING TERPENE COMPOSITIONS FROM WOOD DRYING EXHAUST". World International Patent Office (WIPO) Website. https://patentscope.wipo.int/search/en/detail.jsf?docId=WO2019148208 (Accessed September 18, 2021).

[42] "METHODS AND SYSTEMS FOR CONTROLLING EMISSIONS FROM WOOD DRYING PROCESSES". WIPO Website. https://patentscope.wipo.int/search/en/detail.jsf?docId=WO2019148207 (Accessed September 18, 2021).

[43] "Environmental C&C Inc". Environmental C&C website. https://www.environmentalcc.com/ (Accessed September 18, 2021).

[44] "Emissions Control Technology Captures Emissions to Clean the Air". Captis Aire website. https://www.captisaire.com/ (Accessed November 21, 2021).

[45] "The Trusted Voice of the Clean Air Technology Industry". Institute of Clean Air Companies (ICAC) website. https://www.icac.com/ (Accessed September 18, 2021)

X. ACKNOWLEDGEMENTS:

The author recognizes inputs from a broad spectrum of personnel at the commercial OSB manufacturer and at Environmental C&C[46] including Harold Cowles, John Berger P.E., and Jim Starek.

CAPTIS AIRE CAPTURE-CONCENTRATE-CLEAN	Environmental C&C, Inc. Solvent Recovery & VOC Abatement Systems
Kim TUTIN Captis Aire LLC [47] 1612 W. Cleveland Ave, Suite 001 East Point, GA 30344-3213 Tel: +01 404 5802795, E-mail: kim@captisaire.com Website: www.captisaire.com	Harold COWLES Environmental C&C Inc. 16705 E. Calle de las Estrellas Road Scottsdale, AZ 85262-5982 Tel: +01 518 2250315, E-mail: halc@ecnc.com Website: https://www.environmentalcc.com/

[46] "Environmental C&C Inc". Environmental C&C website. https://www.environmentalcc.com/ (Accessed September 18, 2021).

[47] "Emissions Control Technology Captures Emissions to Clean the Air". Captis Aire website. https://www.captisaire.com/ (Accessed November 21, 2021).

629552 Green Hydrogen – Removing Hurdles and Creating the Right Incentives to Make It Viable

Abstract:

Across the world, strategies are written to support the emerging hydrogen economy. Hydrogen, routinely produced and used as a chemical substance, promises no direct emissions---aside from water vapors---at the point of use. However, due to the imperfect efficiency of electrolyzers, the balance-of-plant costs, and the lack of current transportation infrastructure, it is not yet cost competitive without government incentives. Realizing this, governments have funded research as well as technology demonstration projects and are also implementing changes to the tax code to further incentivize the commercialization.

This talk will consider the primary challenges of green hydrogen implementation and the current and possible future regulatory landscapes that will shape the future adoption of green hydrogen in the US.

628640 A Coating Strategy for Stable H_2 Production from Semiconductor Particles

Abstract:

Efficient semiconductors have been updating the records of photoelectrochemical (PEC) solar-to-hydrogen (STH) conversion efficiencies, owing to the recent breakthrough in protective coatings. The solar energy can be directly stored in hydrogen. I will introduce photocatalytic particles suspended in water or sprayed onto a panel: those particles coevolving two chemicals at their surfaces—the fabrication for these coevolution photocatalysts promising solar conversion at scale. However, the design principles for reaching 25% STH efficiency limit have been elusive. Based on our holistic understanding of the photophysical, electrocatalytic, and transport processes coupled at the nanoscale, we employ a stabilization coating to coevolve H_2 at a record rate of 48.5 $mmol \cdot h^{-1} \cdot g^{-1}$ or 2.5 mL $H_2 \cdot h^{-1} \cdot cm^{-2}$ under 1-sun solar illumination. We also discuss strategies to (1) produce electrochemically pressurized H_2, (2) to design two types of coated particles for z-scheme water splitting, and (3) to achieve H_2/O_2 separation without the generation of explosive mixtures. The PEC hydrogen as an energy carrier can produce electricity or heat, desirable for the urgently needed long-duration renewable energy storage.

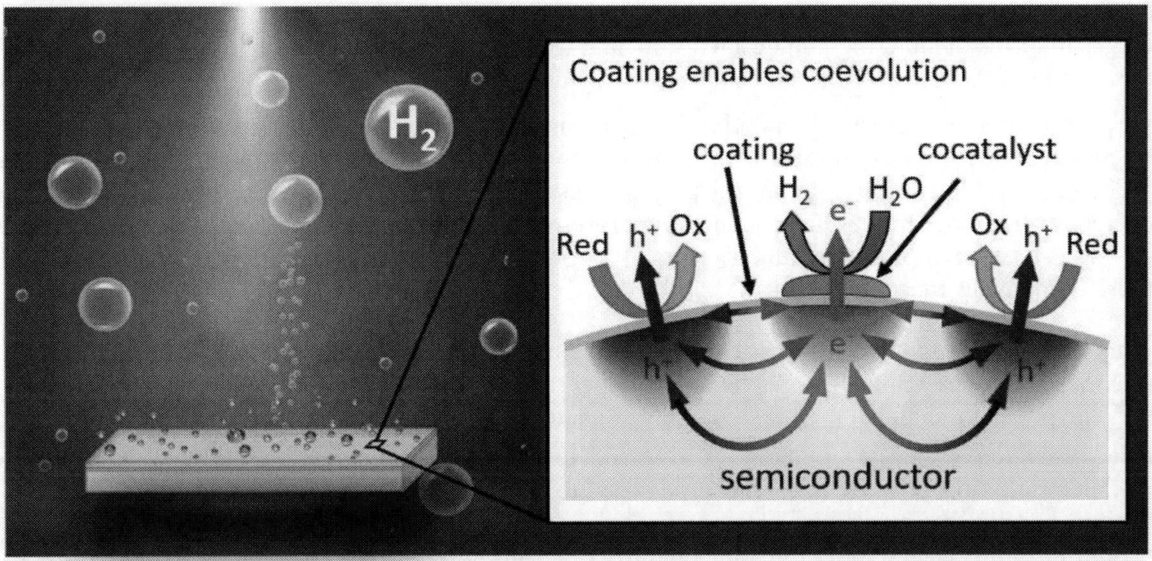

621138 Nickel Ferrite - Ceria Composites for Solar Thermochemical Fuel Production – Efficiency Analysis and Design Goals

Abstract:

Thermochemical redox cycles are a promising pathway for converting solar thermal energy directly to energy carriers like syngas and liquid fuels. Significant efficiency improvements and price reduction are needed before this technology can produce hydrogen at less than \$3/kg-H_2 – a key price target for making green fuels viable. Along with reactor design, identifying better redox materials can have the greatest impact on cycle efficiency. Several redox materials have been investigated in recent years, including doped and undoped ceria, perovskites, ferrites, and multi-phase or composite materials (for example, ceria-perovskite composites[1]). Recent system-level efficiency analyses have shown that undoped ceria is generally more efficient at converting heat to fuels, as compared to zirconia-doped ceria (CZO) and most reported perovskites[2,3]. Even though the CZO and perovskites have higher gravimetric oxygen carrying capacity, they are more efficient than ceria only in a very limited set of operating conditions like isothermal cycling (which is unlikely to be the most efficient mode of operation[4]). The low efficiency of CZO and perovskites is predominantly due to their low conversion during the water-splitting step. This necessitates circulation of large amounts of unreacted steam, and incurs heating and separation energy penalties.

Ferrites can be thermodynamically better suited for fuel production than ceria, but they face other challenges in terms of sintering and deactivation, and low mass diffusivity leading to slow water-splitting kinetics. Some of these problems can be alleviated by using metal doped ferrites (of the form $M_xFe_{3-x}O_4$) and supporting them on refractory materials like zirconia[5]. In the present work, we analyze the thermochemical heat-to-fuel efficiency of the aforementioned material classes and compare them to the efficiency of nickel ferrite-ceria composites. To the best of our knowledge, the latter material has not been considered for thermochemical fuel production before. We identify operating conditions under these ferrite composites can exceed the maximum efficiency of ceria, and in doing so, identify material design goals that these composites must meet.

In this talk we will present our system-level efficiency model for calculating heat-to-fuel efficiency of metal oxides and their composites. Such models depend heavily on the characteristics of the reactor system. The present model is based on the novel 'Reactor Train' system developed by the author and presented elsewhere[6]. The system employs solid and gas-phase heat recovery, and uses waste heat to produce electricity in a power cycle. Vacuum is used to enhance thermal reduction. Thermodynamic equilibrium and realistic mechanical pump efficiencies are used to evaluate pumping and gas separation work. Equilibrium is used to determine the quantity of steam needed to oxidize the redox material, following the 'fixed bed flow' model[3,7]. Using kinetic models for water-splitting would give a more realistic result. However, they are not employed here because accurate models are not available for all materials considered.

The reduction temperature is varied between 1400-1500°C, and oxidation temperature, oxidation extent and reduction pressure are obtained by optimizing for heat-to-fuel efficiency. Oxidation temperature is constrained to be above 700°C to maintain reasonably fast water-splitting kinetics. The materials considered are CeO_2 (ceria), $Ce_{0.85}Zr_{0.15}O_2$ (CZO), $NiFe_2O_4$ (Nickel Ferrite, unsupported), $NiFe_2O_4/ZrO_2$ (Nickel Ferrite, supported on zirconia) and $NiFe_2O_4/CeO_2$ (Nickel Ferrite-ceria composite). The latter two are composite materials with pure phases - i.e. the formation of solution due to cationic exchange is ignored. This is supported by the experiments of Kodama and co-workers, who found no dissolution of iron into the monoclinic zirconia lattice[8]. On the other hand, iron was found to dissolve in cubic Yttria-stabilized zirconia[9]. This system is not considered in the present study because of a lack of thermodynamic data.

We propose nickel ferrite-ceria composites as a means of combining the high oxygen carrying capacity of nickel ferrite with the high-temperature stability and rapid water-splitting kinetics of ceria. Recently, perovskite-ceria composited have been investigated as a means of speeding up the water-splitting step in perovskites[1]. Although faster reactions were observed, the rate was still much slower than ceria. Unlike those perovskites, nickel ferrites do not suffer from low steam conversion at equilibrium, but they must be supported on refractory supports to control agglomeration and particle growth. Zirconia supports have been successfully employed for this purpose[5]. However, zirconia does not participate in redox reactions, and has a low oxygen diffusivity. On the other hand, ceria can serve as an "active support" by participating in redox reactions and contributing to the overall oxygen carrying capacity of the composite. Ceria also has a high oxygen diffusivity, which can be further enhanced with acceptor doping, albeit with some effect on oxygen carrying capacity. Recently, gadolinia-doped ceria was demonstrated as an attractive "active support" material for hematite in chemical looping cycles[10].

Results from our cycle efficiency analysis are summarized in fig. 1. We report that ceria can reach upto 29.5% heat-to-fuel efficiency in the base case (reduction temperature 1500°C). CZO has a much lower optimum efficiency of 24. 5%. Unsupported Nickel ferrite has the highest heat-to-fuel efficiency of 34% at the same reduction temperature, demonstrating its favorable thermodynamic properties. However, such performance is unlikely to be realized in practice because the material would sinter extensively at this temperature. Zirconia supported nickel-ferrites (20% ferrite mass fraction) have been shown to be stable at a reduction temperature of 1400°C[5]. This system has a lower efficiency of 15% due to the lower reduction temperature, as well as the added thermal mass of zirconia. Even this efficiency might be difficult to achieve in practice because it needs an oxidation temperature of 700°C, where water-splitting kinetics may be too slow. Increasing the oxidation temperature to 1000°C[5] further reduces efficiency to below 10%. We note that ceria is shown to be stable at a reduction temperature of 1500°C[11].

The Nickel ferrite-ceria composite with 50 wt% ferrite has an optimum efficiency of 33% at a reduction temperature of 1500°C and oxidation temperature of 700°C. Decreasing the ferrite mass fraction to 20% reduces the optimum efficiency to 32%. Both these values exceed the base case efficiency of ceria (see fig. 1). However, this advantage is lost when the composite is reduced 1400°C. The composite also becomes less efficient when the oxidation temperature is increased to 1000°C. At this temperature, the water-splitting reaction becomes thermodynamically less favorable. The variation of heat-to-fuel efficiency with oxidation temperature is shown in fig. 2. However, there is reason to believe that the nickel ferrite-ceria composite can split water at low temperatures - pure ceria was observed to oxidize rapidly in steam at temperatures as low as 600°C [5]. Moreover, the composite material is less sensitive to a drop in solid heat recovery effectiveness than ceria, as shown in fig. 1. This is because of the waste heat recovery system that produces electricity, and the favorable thermodynamics of Nickel ferrite.

We thus conclude that although nickel ferrite has better thermodynamic performance, practical considerations make it unattractive compared to ceria, which is the current state-of-the-art material. In view of this, nickel ferrite-ceria composites emerge as promising redox materials with the potential to surpass ceria. However, they must be designed to withstand a reduction temperature of 1500°C, and achieve reasonably fast water-splitting below 800°C. Their efficiency can be increased further by increasing the weight fraction of ferrite in the composite, if this can be achieved without negatively impacting material stability and water-splitting kinetics. Experimental work on the synthesis, characterization and thermochemical testing of these composites is ongoing.

1. Bork *et al.* $La_{0.65}Sr_{0.35}MnO_3$-CeO_2 Composites for Solar Thermochemical Fuel Production. *ACS Appl. Mater. Interfaces* **12**, 32622–32632 (2020).
2. Muhich *et al.* Comparing the solar-to-fuel energy conversion efficiency of ceria and perovskite based thermochemical redox cycles for splitting H2O and CO2. *Int. J. Hydrogen Energy* **43**, 18814–18831 (2018).

3. Li, S. *et al.* Thermodynamic Guiding Principles for Designing Nonstoichiometric Redox Materials for Solar Thermochemical Fuel Production: Ceria, Perovskites, and Beyond. *Energy Technol.* **2000925**, 1–18 (2021).
4. Hathaway *et al.* Demonstration of a Solar Reactor for Carbon Dioxide Splitting via the Isothermal Ceria Redox Cycle and Practical Implications. *Energy and Fuels* **30**, 6654–6661 (2016).
5. Kodama *et al.* Comparison Studies Of Reactivity On Nickel-Ferrite And Cerium oxide Redox Materials For Two Step Thermochemical Water Splitting Below 1400°C. 1–7 (2011).
6. Patankar *et al.* A Reactor Train System for Efficient Solar Thermochemical Fuel Production. *In preparation* (2021).
7. Bulfin *et al.* Thermodynamics of CeO_2 thermochemical fuel production. *Energy and Fuels* **29**, 1001–1009 (2015).
8. Gokon *et al.* T. Monoclinic zirconia-supported Fe3O4 for the two-step water-splitting thermochemical cycle at high thermal reduction temperatures of 1400-1600 °C. *Int. J. Hydrogen Energy* **34**, 1208–1217 (2009).
9. Coker *et al.* Compositional and operational impacts on the thermochemical reduction of CO2to CO by iron oxide/yttria-stabilized zirconia. *RSC Adv.* **11**, 1493–1502 (2021).
10. Zeng *et al.* Iron oxides with gadolinium-doped cerium oxides as active supports for chemical looping hydrogen production. *Chem. Eng. J.* **396**, 125153 (2020).
11. Marxer *et al.* Solar thermochemical splitting of CO_2 into separate streams of CO and O_2 with high selectivity, stability, conversion, and efficiency. *Energy Environ. Sci.* **10**, 1142–1149 (2017).

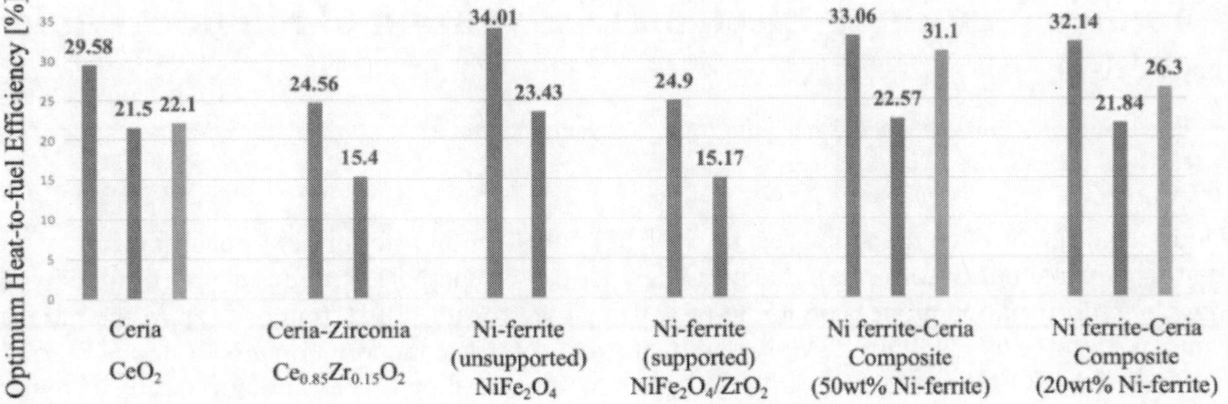

Figure 1. Optimum heat-to-fuel efficiencies of the six material compositions considered (left to right: ceria, Zr-doped ceria, $NiFe_2O_4$, 20 wt% supported $NiFe_2O_4$ on zirconia, $NiFe_2O_4$-ceria composite with 50 wt% ferrite, and 20 wt% ferrite). Colors of bars represent the following:

Red: $T_{reduction}$ = 1500°C, Solid Heat recovery = 80% (Base case)

Blue: $T_{reduction}$ = **1400°C**, Solid Heat recovery = 80%

Green: $T_{reduction}$ = 1500°C, Solid Heat recovery = **50%**

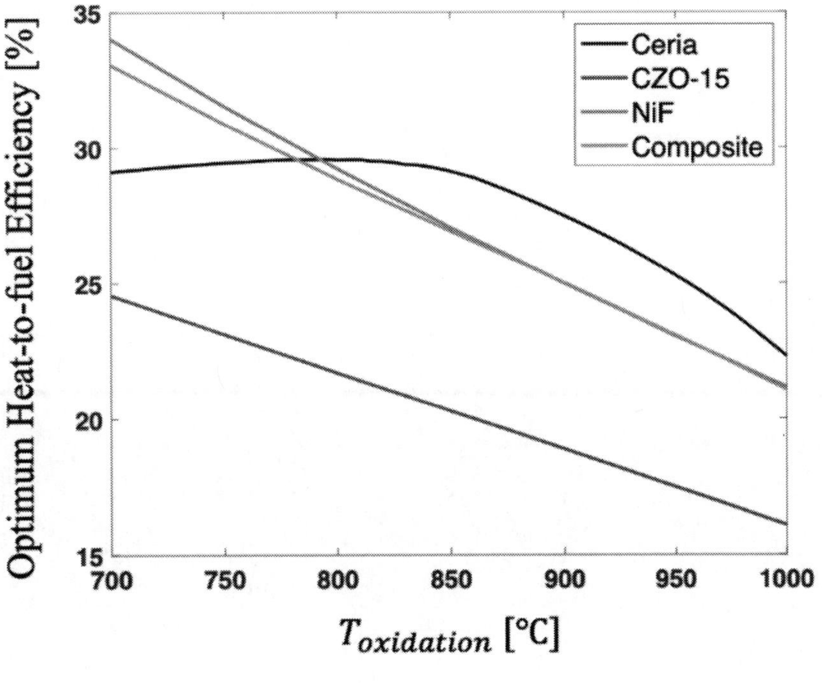

Figure 2. Variation of base case heat-to-fuel efficiency of CeO_2 (ceria - black), $Ce_{0.85}Zr_{0.15}O_2$ (Zirconia-doped ceria - blue), $NiFe_2O_4$ (unsupported Nickel Ferrite - red), and $NiFe_2O_4/CeO_2$ (Nickel ferrite ceria composite, 50 wt% ferrite – magenta).

621140 Boosting Electroreduction Kinetics of Nitrogen to Ammonia Viatuning Electron Distribution of Single-Atomic Iron Sites

Abstract:

Electrocatalytic nitrogen reduction reaction (NRR) plays a vital role for next-generation electrochemical energy conversion technologies. However, the NRR kinetics is still limited by the sluggish hydrogenation process on noble-metal-free electrocatalyst. Herein, we report the rational design and synthesis of a hybrid catalyst with atomic iron sites anchored on a N,O-doped porous carbon (FeSA-NO-C) matrix of an inverse opal structure, leading to a remarkably high NH_3 yield rate of 31.9 $\mu gNH_3\ h^{-1}\ mg^{-1}{}_{cat}$. and Faradaic efficiency of 11.8% at -0.4 V for NRR electrocatalysis, outperformed almost all previously reported atomically dispersed metal-nitrogen-carbon catalysts. Theoretical calculations revealed that the observed high NRR catalytic activity for the FeSA-NO-C catalyst stemmed mainly from the optimized charge-transfer between the adjacent O and Fe atoms homogenously distributed on the porous carbon support, which could not only significantly facilitate the transportation of N_2 and ions but also effectively decrease the binding energy between the isolated Fe atom and *N2 intermediate and the thermodynamic Gibbs free energy of the rate-determining step ($*N_2 \rightarrow *NNH$).

632297 An Electrofuel Revolution: How Direct Carbon Reduction Can Electrify the Future

Abstract:

While solar, wind, and other renewable energy sources promise affordable electricity with minimal carbon emissions, electricity cannot replace all of the functions of liquid fuels. Liquid fuels remain the most effective means to store energy for significant durations at small to medium scale. Therefore, there remains a need for liquid fuels but ones without a significant carbon footprint. One such route, whether through chemical, electrochemical, or biochemical means, is to use carbon dioxide (CO_2) as the feedstock to produce those fuels rather than petroleum. Historically, CO_2 upgrading has been performed in capital-intensive, large-scale reverse water gas shift reactors to generate more functional carbon monoxide for further processing. Recently, however, researchers have begun investigating electrocatalysts that can convert carbon dioxide to valuable fuel additives and chemical precursors directly at, or near, ambient conditions with reduced emissions and capital requirements. Unlike hydrogen, ammonia, or batteries, electrofuels, and other carbon-neutral fuels, can serve as drop-in replacements to conventional petrochemical fuels without significant change to the end-use infrastructure.

This talk will highlight the opportunities and challenges for electrofuel production, and how the evolving regulatory environment may incentivize widespread adoption.

622429 Engineering Yeast to Synthesize Semiconductor Nanoparticles Allowing Light Driven CO_2 Fixation and Biofuel Production

Abstract:

Engineered photosynthetic systems aim to capture solar energy and reduce carbon dioxide. These systems use light to create conditions favorable for net carbon fixation to produce biomass and/or liquid fuels. We describe a hybrid inorganic-biological system that combines the light harvesting properties of a semiconductor system that when combined with genetic engineering can alter yeast cell redox state and favor generation of useful products. Here we show this system can be used to increase ethanol production, a common biofuel, through reductive carboxylation stimulated by biologically produced cadmium sulfide nanoparticles and light. This illustrates how use of this system can alter yeast metabolism and allow production of many metabolites.

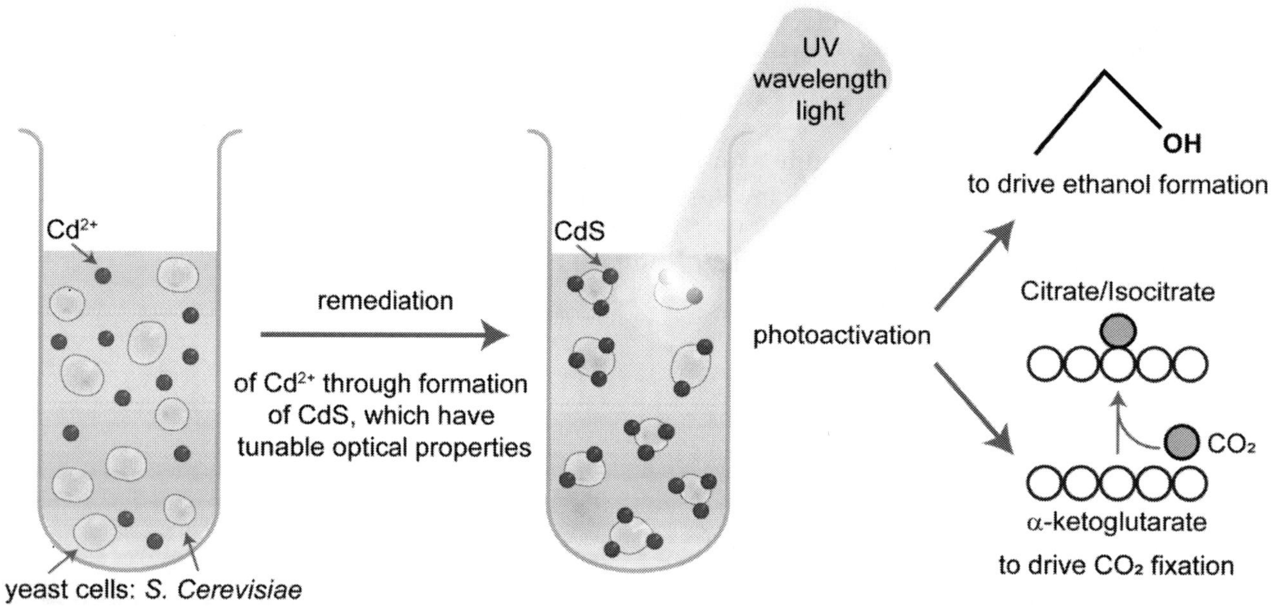

623407 Direct Conversion of Industrial Exhaust Gas to a C40 Natural Product By Biologic/Inorganic Hybrid Catalysis

Abstract:

Over one-third CO_2 emission arises from industry, and valorization of CO_2 in industrial exhaust gas is key to establishing a carbon-neutral society. However, there has been few reports on the conversion of CO_2 from real industrial exhaust gas to complex and value-added chemicals. Although electrochemical CO_2 reduction has been attracted increasing attention and great progress has been achieved, the product profiles based on inorganic catalysis are only limited to a few simple chemicals, such as formate, acetate, and ethanol. Optionally, biocatalysis can convert CO_2 to more complex chemicals, while the supply of reducing power remains a grand challenge. Therefore, the coupling of inorganic electrochemical catalysis and biocatalysis offers a promising route for powering sustainable CO_2 conversion with renewable electricity. Herein, we successfully achieved the production of lycopene using real exhaust gas from a coal-fired power plant for the first time, which represented the most complex nonnative molecules from CO_2 reduction. By coupling *de novo* lycopene biosynthesis with water electrolysis, a microbial electrosynthesis (MES) system was developed. Taking advantage of this biologic/inorganic hybrid catalysis, lycopene was produced with CO_2 as the sole carbon source and H_2 derived from electrochemical water splitting as the reducing power. Moreover, the produced lycopene provided long-term cytoprotective capacity against reactive oxygen species, which addressed the incompatibility between biological and electrochemical catalysis. The maximum yield of lycopene was 1.73 mg/L when real exhaust gas was used as feedstock, which indicated the developed MES system was sustainable and robust, and with great potential for practical application in CO_2 valorization. The present study opens a possible route to turn trash to cash for industrial exhaust gas.

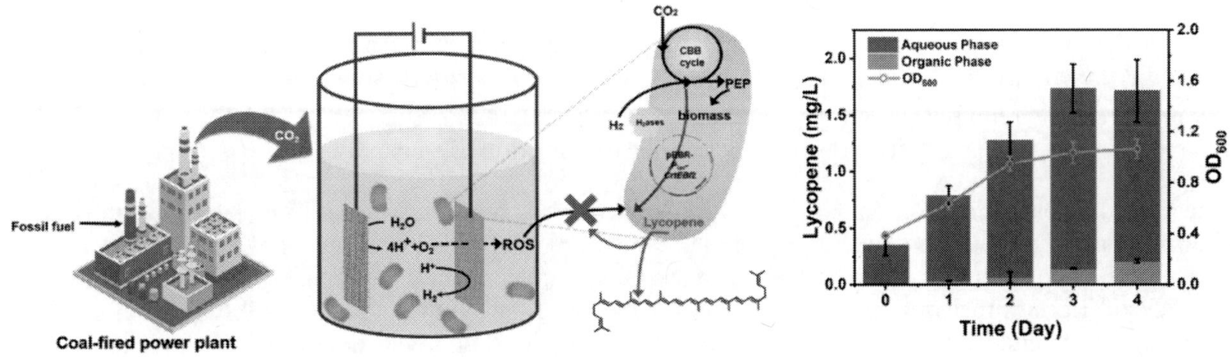

623575 A Comparative Assessment Framework for Sustainable Production of Fuels and Chemicals Explicitly Accounting for Intermittency

Abstract:

The combustion of fossil fuels in the transport sector contributes to anthropogenic CO_2 emissions and climate change. Sustainable fuel concepts are needed to replace established fuels and move toward a sustainable transport sector. Ways to decarbonise transport are electric vehicles (EVs), hydrogen fuel cell vehicles (HFCVs), biofuels, and carbon-neutral synthetic fuels (CNSF). While EVs are making strong advances in short-distance and personal transport, CNSFs may be needed for decarbonising long-distance and heavy goods vehicles. Among CNSFs, carbon-based fuels such as methanol have received considerable attention. However, in a post-fossil economy the prevalence of carbon is not a given. Nitrogen-based fuels like ammonia should be carefully examined. Not only is CO_2 more dilute in air, its reduction is more energetically expensive compared to that of N_2. Furthermore, sustainable fuels are required to be produced using renewable energy. When the energy source of the process is intermittent, as is the case with wind, solar, and tidal energy, intermittency needs to be taken into account in the design of the process. When the analysis of sustainable fuels only includes the synthesis plant assuming fixed cost and availability for the building blocks, the cost of intermittency is neglected and importance of flexibility in processes is ignored.

We developed a framework for the comparison of sustainable fuel production from air, water, and renewable energy. We perform simultaneous design and scheduling of sustainable fuel production including renewable energy source, air separation, electrolysers, and synthesis. All power requirements have to be met by the renewable energy source; heat requirements can be met via heat integration, residual heat requirements are met using hydrogen. In our work, we compare solar methanol and ammonia in London and Dubai with regard to process design and operation as well as cost.

Our results show that the optimal route depends on the location. In London, where the solar pattern exhibits high seasonality, the route requiring the least amount of gas storage is optimal. In Dubai, with lower seasonality, the most energy efficient route, requiring the least amount of PV and electrolyser capacity, is the optimal choice. Figure 2 shows the optimal routes for methanol.

We further determined that nitrogen-based fuels may be competitive with carbon-based fuels. When taking into account the energy needed to separate CO_2 or N_2 from air, it becomes apparent that the production of nitrogen-based fuels requires less energy and hence may be less costly. Our results support that carbon's dominance in fuels may not prevail throughout the decarbonisation of transport.

We quantify the cost of intermittency to be as high as two thirds of the total cost. Figure 3 shows the cost of intermittency in London and Dubai for methanol and ammonia. This implicates that for CNSF, specifically solar fuels, the intermittency needs to be taken into account when making cost predictions. The value of flexibility of processes is accordingly high. Operating the synthesis plant flexibly reduces the seasonal storage needed and with it the overall cost. This implies that novel fuel production processes, such as one-step synthesis like photocatalytic ammonia production, or more flexible processes, may have advantages over established production routes even if they have higher individual CAPEX.

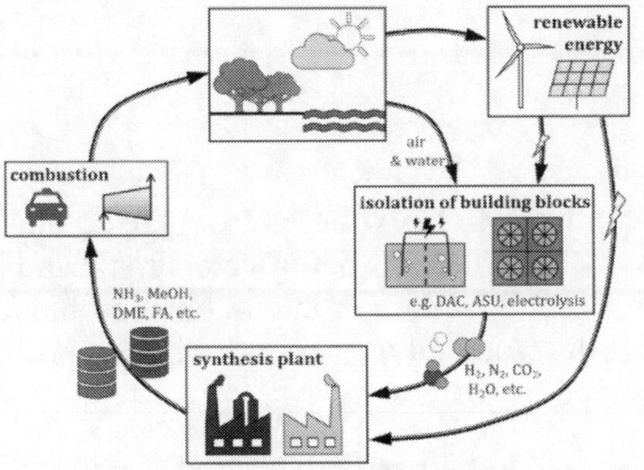

Figure 1: Circular economy for fuels.

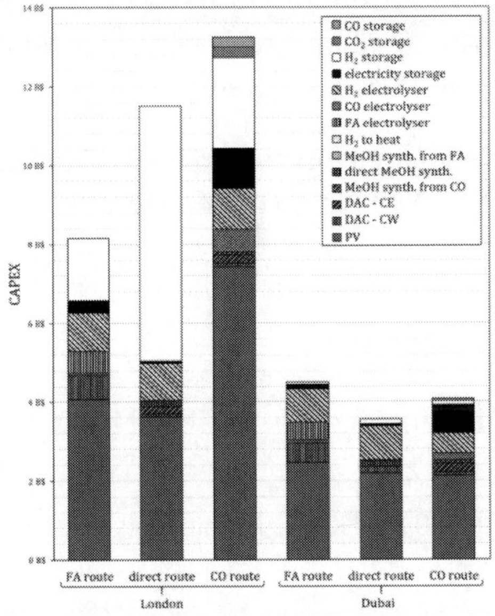

Figure 2: Comparison of methanol production routes via formic acid (FA) and carbon monoxide (CO) as well as direct route in London and Dubai.

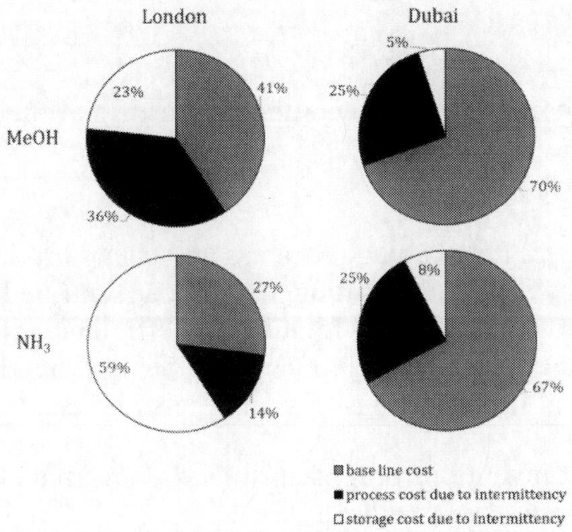

Figure 3: Cost of intermittency.

620281 Cheap Energy Is a Key to Competitive Artificial Fuels

Abstract:

Cheap energy is a key to competitive artificial fuels

Astract

Synthetic & renewable fuels mass and cheap production from CO2 extracted from the air and H2 created by the water splitting, e.g. by concentrated rays of magnetrons being very simple and cheap electronic device of very high energy efficiency near 100%, it is not a problem, but a problem to have a electricity cheaper USD 0.01/kWh, being around the clock and all year round. The last problem was my special investigations of some years and can be solved by different ways.

Different inventions promise primary renewable electricity cheaper USD 0.01/kWh each at 5 years payback and then practically free.
They are.
1. Solar thermal heterodyne. 2020. It allows a rectenna to be feasible at present level of rectifiers.
2. VAWT (vertical axis wind turbine) comprising a chain of units
rotated oppositely to neighboring ones, 2016.
3. Even well know solar PV panels can harvest 65% more solar energy with a help of a planar foil reflector, adding additional sun to the PV panel, 2020. Now such panels are dramatically cheapened,

and other energy technologies.

For example.

4. Sailboats with hydro-generator. 2017. Its electricity is transformed on-board in hydrogen, and then in more transportable fuels

(Fresh look on Wind Power Generation – Upgraded, 25/02/2019, David Judbarovski,

2019 AIChE Annual Meeting (ISBN: 978-0-8169-1112-7),
https://www.aiche.org/proceedings/people/david-judbarovski }

Breeding investment allows by credit for a first power plant in five steps to invest and build some tens analogous power plants. .

So OPEX is negligible vs. CAPEX, can be full cost of artificial intermediary fuel for 100% off-grid electricity by 2.5 cent/kWh. For housing such autonomy can be provided by local fuel-in-electricity substations. Autonomous supply can be paralleled by much cheaper direct grid supply.

CO2 can be USD 15.0/ton being my estimation with a good reserve, and being produced in quantity of 320 ton/m2/year in production ponds and by electricity cheaper than USD 0.01/kWh

(579f) CO2 Mass Production from the Air Very Cheaply, 2020 Virtual AIChE Annual Meeting (ISBN: 978-0-8169-1114-1), https://www.aiche.org/proceedings/people/david-judbarovski

AUTHOR INDEX